# CALCULATIONS FOR GCSE CHEMISTRY

*Other Chemistry books for schools from Stanley Thornes (all available from Stanley Thornes, Cheltenham):*

A-LEVEL CHEMISTRY by E N Ramsden
CALCULATIONS FOR A-LEVEL CHEMISTRY Third Edition by E N Ramsden
A FIRST CHEMISTRY COURSE by E N Ramsden
CHEMISTRY TESTS FOR FIRST EXAMINATIONS by A Porter and T Wood
MODERN ORGANIC CHEMISTRY Third Edition by A Atkinson
A MODERN APPROACH TO COMPREHENSIVE CHEMISTRY Third Edition by G N Gilmore

# CALCULATIONS FOR GCSE CHEMISTRY

**E N Ramsden BSc, PhD, DPhil**

**Stanley Thornes (Publishers) Ltd**

First published in 1987 by:
Stanley Thornes (Publishers) Ltd
Ellenborough House
Wellington Street
CHELTENHAM GL50 1YD
England

Reprinted 1988, 1989, 1992, 1993

British Library Cataloguing in Publication Data
Ramsden, E. N.
Calculations for GCSE chemistry
1. Chemistry—Mathematics
I. Title
540'.1'51 QD39.3.M3

ISBN 0–85950–667–3

Typeset in Palatino by KEYTEC, Bridport, Dorset
Printed and bound in Great Britain at The Bath Press, Avon

# Contents

# Preface

This book is a second edition of *Calculations for O-level Chemistry*. The content has been altered to fit the requirements of the new GCSE syllabuses. All the types of numerical problems listed by the six GCSE Examination Boards have been covered. A brief treatment of the theoretical background to each type of problem is given, and is followed by a series of worked examples. The problems are divided into three sections of increasing difficulty. The arithmetic in Sections 1 and 2 has been kept simple by basing the problems on a selection of compounds with relative molecular masses which are round numbers, such as $NH_4NO_3 = 80$, $MgSO_4 = 120$, $CaBr_2 = 200$. Each Section 3 consists of longer questions and more difficult questions. After the first GCSE examinations have been set, I hope to include some questions from past papers in these sections. If a pupil has difficulty with a problem, he or she can return for help to the theoretical section and to the worked examples. Thus the book can be used for private study as well as for class work.

The concept of the mole is the thread which knits together the calculations on reacting masses of solids, reacting volumes of gases, empirical formulae, reacting volumes of solutions and heats of reaction. The pupil learns to look at the equation for the reaction and ask himself or herself how many moles of reactant are involved.

The Association for Science Education publication, *Chemical Nomenclature, Symbols and Terminology for use in School Science* (Third Edition, 1985) has been followed in matters of terminology.

The Examination Boards differ in their usage, for example between litres and cubic decimetres. Some boards require candidates to appreciate that mole is the unit of amount of substance; others do not demand so precise a treatment. Some boards do not require every type of calculation included in this book, for example, calculations on heat of reaction and calculations on titration are not required by every board. Readers should make sure that they know exactly what is included in the syllabus they are following.

E N Ramsden
Hull, 1987

# Acknowledgements

I would like to repeat my thanks to all those who helped me during the preparation of the first edition of this book, especially Chris Baker, who suggested the project, and Stephanie Cox, who checked the answers. I thank the teachers who have given me their views on the book and the pupils who have helped me with their constructive comments. I thank Stanley Thornes (Publishers) for the clear layout of the text and the care that has gone into both editions of the book. Finally, I thank my family for their support and encouragement.

E.N.R.

# 1. Formulae and Equations

Calculations are based on formulae and on equations. In order to tackle the calculations in this book, you will have to be quite sure you can work out the formulae of compounds correctly, and that you can balance equations. This section is a revision of work on formulae and equations.

## Formulae

Electrovalent compounds consist of oppositely charged ions. The compound formed is neutral because the charge on the positive ion (or ions) is equal to the charge on the negative ion (or ions). In sodium chloride, NaCl, one sodium ion, $Na^+$, is balanced in charge by one chloride ion, $Cl^-$.

*This is how the formulae of electrovalent compounds can be worked out*

| | |
|---|---|
| *Compound* | *Zinc chloride* |
| Ions present are | $Zn^{2+}$ and $Cl^-$ |
| Now balance the charges | One $Zn^{2+}$ ion needs two $Cl^-$ ions |
| Ions needed are | $Zn^{2+}$ and $2Cl^-$ |
| The formula is | $ZnCl_2$ |
| *Compound* | *Sodium sulphate* |
| Ions present are | $Na^+$ and $SO_4^{2-}$ |
| Now balance the charges | Two $Na^+$ balance one $SO_4^{2-}$ |
| Ions needed are | $2Na^+$ and $SO_4^{2-}$ |
| The formula is | $Na_2SO_4$ |
| *Compound* | *Aluminium sulphate* |
| Ions present are | $Al^{3+}$ and $SO_4^{2-}$ |
| Now balance the charges | Two $Al^{3+}$ balance three $SO_4^{2-}$ |
| Ions needed are | $2Al^{3+}$ and $3SO_4^{2-}$ |
| The formula is | $Al_2(SO_4)_3$ |
| *Compound* | *Iron (II) sulphate* |
| Ions present are | $Fe^{2+}$ and $SO_4^{2-}$ |
| Now balance the charges | One $Fe^{2+}$ balances one $SO_4^{2-}$ |
| Ions needed are | $Fe^{2+}$ and $SO_4^{2-}$ |
| The formula is | $FeSO_4$ |
| *Compound* | *Iron (III) sulphate* |
| Ions present are | $Fe^{3+}$ and $SO_4^{2-}$ |
| Now balance the charges | Two $Fe^{3+}$ balance three $SO_4^{2-}$ |
| Ions needed are | $2Fe^{3+}$ and $3SO_4^{2-}$ |
| The formula is | $Fe_2(SO_4)_3$ |

You need to know the charges of the ions in Table 1.1. Then you can work out the formula of any electrovalent compound.

You will notice that the compounds of iron are named iron(II) sulphate and iron(III) sulphate to show which of its valencies iron is using in the compound. This is always done with the compounds of elements of variable valency.

Table 1.1 *Symbols and valencies of common ions*

| Name | Symbol | Valency | Name | Symbol | Valency |
|---|---|---|---|---|---|
| Hydrogen | $H^+$ | 1 | Hydroxide | $OH^-$ | 1 |
| Ammonium | $NH_4^+$ | 1 | Nitrate | $NO_3^-$ | 1 |
| Potassium | $K^+$ | 1 | Chloride | $Cl^-$ | 1 |
| Sodium | $Na^+$ | 1 | Bromide | $Br^-$ | 1 |
| Silver | $Ag^+$ | 1 | Iodide | $I^-$ | 1 |
| Copper(I) | $Cu^+$ | 1 | Hydrogen-carbonate | $HCO_3^-$ | 1 |
| Barium | $Ba^{2+}$ | 2 | Oxide | $O^{2-}$ | 2 |
| Calcium | $Ca^{2+}$ | 2 | Sulphide | $S^{2-}$ | 2 |
| Copper(II) | $Cu^{2+}$ | 2 | Sulphite | $SO_3^{2-}$ | 2 |
| Iron(II) | $Fe^{2+}$ | 2 | Sulphate | $SO_4^{2-}$ | 2 |
| Lead | $Pb^{2+}$ | 2 | Carbonate | $CO_3^{2-}$ | 2 |
| Magnesium | $Mg^{2+}$ | 2 | | | |
| Zinc | $Zn^{2+}$ | 2 | | | |
| Aluminium | $Al^{3+}$ | 3 | Phosphate | $PO_4^{3-}$ | 3 |
| Iron(III) | $Fe^{3+}$ | 3 | | | |

# Equations

Having symbols for elements and formulae for compounds gives us a way of representing chemical reactions.

**Example 1** Instead of writing 'Copper(II) carbonate forms copper(II) oxide and carbon dioxide', we can write

$$CuCO_3 \rightarrow CuO + CO_2$$

The atoms we finish with are the same in number and kind as the atoms we start with. We start with one atom of copper, one atom of carbon and three atoms of oxygen, and we finish with the same. This makes the two sides of the expression equal, and we call it an **equation**. A simple way of conveying a lot more information is to include **state symbols** in the equation. These are (s) = solid, (l) = liquid, (g) = gas, (aq) = in solution in water. The equation

$$CuCO_3(s) \rightarrow CuO(s) + CO_2(g)$$

tells you that solid copper(II) carbonate dissociates to form solid copper(II) oxide and the gas carbon dioxide.

**Example 2** The equation

$$Zn(s) + H_2SO_4(aq) \rightarrow ZnSO_4(aq) + H_2(g)$$

tells you that solid zinc reacts with a solution of sulphuric acid to give a solution of zinc sulphate and hydrogen gas. Hydrogen is written as $H_2$, since each molecule of hydrogen gas contains two atoms.

**Example 3** Sodium carbonate reacts with dilute hydrochloric acid to give carbon dioxide and a solution of sodium chloride. The equation could be

$$Na_2CO_3(s) + HCl(aq) \rightarrow CO_2(g) + NaCl(aq) + H_2O(l)$$

but, when you add up the atoms on the right, you find that they are not equal to the atoms on the left. The equation is not '*balanced*', so the next step is to *balance* it. Multiplying NaCl by two gives

$$Na_2CO_3(s) + HCl(aq) \rightarrow CO_2(g) + 2NaCl(aq) + H_2O(l)$$

This makes the number of sodium atoms on the right-hand side equal to the number on the left-hand side. But there are two chlorine atoms on the right-hand side, therefore the HCl must be multiplied by two:

$$Na_2CO_3(s) + 2HCl(aq) \rightarrow CO_2(g) + 2NaCl(aq) + H_2O(l)$$

The equation is now **balanced**.

When you are balancing a chemical equation, the only way you do it is to multiply the number of atoms or molecules. You never try to alter a formula. In the above example, you got two chlorine atoms by multiplying HCl by two, not by altering the formula to $HCl_2$, which does not exist.

**The steps in writing an equation are**

1. Write a word equation
2. Put in the symbols and formulae (symbols for elements, formulae for compounds and state symbols)
3. Balance the equation

**Example 4** The reaction between sodium and water to form hydrogen and sodium hydroxide solution. Work through the three steps:

1. Sodium + Water → Hydrogen + Sodium hydroxide solution
2. $Na(s) + H_2O(l) \rightarrow H_2(g) + NaOH(aq)$
3. $2Na(s) + 2H_2O(l) \rightarrow H_2(g) + 2NaOH(aq)$

**Example 5** When methane burns,

Methane + Oxygen → Carbon dioxide and Water:

$$CH_4(g) + O_2(g) \rightarrow CO_2(g) + H_2O(g)$$

There is one carbon atom on the left-hand side and one carbon atom on the right-hand side. There are four hydrogen atoms on the left-hand side, and therefore we need to put four hydrogen atoms on the right-hand side. Putting $2H_2O$ on the right-hand side will accomplish this:

$$CH_4(g) + O_2(g) \rightarrow CO_2(g) + 2H_2O(g)$$

There is one molecule of $O_2$ on the left-hand side and four O atoms on the right-hand side. We can make the two sides equal by putting $2O_2$ on the left-hand side:

$$CH_4(g) + 2O_2(g) \rightarrow CO_2(g) + 2H_2O(g)$$

This is a balanced equation. The numbers of atoms of carbon, hydrogen and oxygen on the left-hand side are equal to the numbers of atoms of carbon, hydrogen and oxygen on the right-hand side.

**Example 6** When propane burns,

Propane + Oxygen → Carbon dioxide and Water:

$$C_3H_8(g) + O_2(g) \rightarrow CO_2(g) + H_2O(g)$$

As there are three C atoms on the left-hand side, there must be $3CO_2$ molecules on the right-hand side:

$$C_3H_8(g) + O_2(g) \rightarrow 3CO_2(g) + H_2O(g)$$

As there are eight H atoms on the left-hand side, there must be $4H_2O$ on the right-hand side:

$$C_3H_8(g) + O_2(g) \rightarrow 3CO_2(g) + 4H_2O(g)$$

Counting the oxygen atoms, there are two on the left-hand side and ten on the right-hand side. Putting $5O_2$ on the left-hand side will make the two sides equal:

$$C_3H_8(g) + 5O_2(g) \rightarrow 3CO_2(g) + 4H_2O(g)$$

This is a balanced equation.

## Practice with equations

1. For practice, try writing the equations for the reactions:
   (a) Hydrogen + Copper oxide → Copper + Water
   (b) Carbon + Carbon dioxide → Carbon monoxide
   (c) Carbon + Oxygen → Carbon dioxide
   (d) Magnesium + Sulphuric acid → Hydrogen + Magnesium sulphate
   (e) Copper + Chlorine → Copper(II) chloride

2. Now try writing balanced equations for the reactions:
   (a) Calcium + Water → Hydrogen + Calcium hydroxide solution
   (b) Copper + Oxygen → Copper(II) oxide
   (c) Sodium + Oxygen → Sodium oxide
   (d) Iron + Hydrochloric acid → Iron(II) chloride solution
   (e) Iron + Chlorine → Iron(III) chloride

3. Balance these equations:
   (a) $Na_2O(s) + H_2O(l) \rightarrow NaOH(aq)$
   (b) $KClO_3(s) \rightarrow KCl(s) + O_2(g)$
   (c) $H_2O_2(aq) \rightarrow H_2O(l) + O_2(g)$
   (d) $Fe(s) + O_2(g) \rightarrow Fe_3O_4(s)$
   (e) $Mg(s) + N_2(g) \rightarrow Mg_3N_2(s)$
   (f) $NH_3(g) + O_2(g) \rightarrow N_2(s) + H_2O(g)$
   (g) $Fe(s) + H_2O(g) \rightarrow Fe_3O_4(s) + H_2(g)$
   (h) $H_2S(g) + O_2(g) \rightarrow H_2O(g) + SO_2(g)$
   (i) $H_2S(g) + SO_2(g) \rightarrow H_2O(l) + S(s)$

# 2. Relative Atomic Mass; Relative Molecular Mass; Percentage Composition

## Relative Atomic Mass

Atoms are tiny: one atom of hydrogen has a mass of $1.66 \times 10^{-24}$ g; one atom of carbon has a mass of $1.99 \times 10^{-23}$ g. Numbers as small as this are awkward to handle, and, instead of the actual masses, we use relative atomic masses. Since hydrogen atoms are the smallest of all atoms, one atom of hydrogen was taken as the mass with which all other atoms would be compared. Then,

$$\text{Relative atomic mass} = \frac{\text{Mass of one atom of the element}}{\text{Mass of one atom of hydrogen}}$$

Thus, on this scale, the relative atomic mass of hydrogen is 1, and, since one atom of carbon is 12 times as heavy as one atom of hydrogen, the relative atomic mass of carbon is 12.

The modern method of finding relative atomic masses is to use an instrument called a mass spectrometer. The most accurate measurements are made with volatile compounds of carbon, and it was therefore convenient to change the standard of reference to carbon. There are three isotopes of carbon. **Isotopes** are forms of an element whose atoms have the same number of protons and electrons but have different numbers of neutrons, and therefore different masses. It was decided to use the most plentiful carbon isotope, carbon-12. Thus,

$$\text{Relative atomic mass} = \frac{\text{Mass of one atom of an element}}{(1/12)\ \text{Mass of one atom of carbon-12}}$$

On this scale, carbon-12 has a relative atomic mass of 12.000 000, and hydrogen has a relative atomic mass 1.007 97. Since relative atomic masses are ratios of two masses, they have no units. As this value for hydrogen is very close to one, the value of H = 1 is used in most calculations. A table of approximate relative atomic masses is given on page 85. **The symbol for relative atomic mass is $A_r$.**

## Relative Molecular Mass

A molecule consists of a combination of atoms. You can find the mass of a molecule by adding up the masses of all the atoms in it. You can find the relative molecular mass of a compound by adding the relative atomic masses of all the atoms in a molecule of the compound. For example, you can work out the relative molecular mass of carbon dioxide as follows:

*The formula is* $CO_2$
1 atom of C, relative atomic mass 12 = 12
2 atoms of O, relative atomic mass 16 = 32
Total = 44
Relative molecular mass of $CO_2$ = 44

**The symbol for relative molecular mass is** $M_r$.

A vast number of compounds consist of ions, not molecules. The compound sodium chloride, for example, consists of sodium ions and chloride ions. You cannot correctly refer to a 'molecule of sodium chloride'. For ionic compounds, the term **formula unit** is used to describe the ions which make up the compound. A formula unit of sodium chloride is NaCl. A formula unit of copper(II) sulphate-5-water is $CuSO_4 \cdot 5H_2O$. It is still correct to use the term relative molecular mass for ionic compounds:

$$\text{Relative molecular mass} = \frac{\text{Mass of one formula unit}}{(1/12)\ \text{Mass of one atom of carbon-12}}$$

Thus, for sodium chloride:

*The formula is NaCl*
1 atom of Na, relative atomic mass 23 = 23
1 atom of Cl, relative atomic mass 35.5 = 35.5
Total = 58.5
Relative molecular mass of NaCl = 58.5

We work out the relative molecular mass of calcium chloride as follows:

*The formula is* $CaCl_2$
1 atom of Ca, relative atomic mass 40 = 40
2 atoms of Cl, relative atomic mass 35.5 = 71
Total = 111
Relative molecular mass of $CaCl_2$ = 111

We work out the relative molecular mass of aluminium sulphate as follows:

*The formula is* $Al_2(SO_4)_3$
2 atoms of Al, relative atomic mass 27 = 54
3 atoms of S, relative atomic mass 32 = 96
12 atoms of O, relative atomic mass 16 = 192
Total = 342
Relative molecular mass of $Al_2(SO_4)_3$ = 342

## Problems on Relative Molecular Mass

Work out the relative molecular masses of these compounds:

| | | |
|---|---|---|
| $SO_2$ | $NaOH$ | $KNO_3$ |
| $MgCO_3$ | $PbCl_2$ | $MgCl_2$ |
| $Mg(NO_3)_2$ | $Zn(OH)_2$ | $ZnSO_4$ |
| $H_2SO_4$ | $HNO_3$ | $MgSO_4 \cdot 7H_2O$ |
| $CaSO_4$ | $Pb_3O_4$ | $P_2O_5$ |
| $Na_2CO_3$ | $Ca(OH)_2$ | $CuCO_3$ |
| $CuSO_4$ | $Ca(HCO_3)_2$ | $CuSO_4 \cdot 5H_2O$ |
| $Fe_2(SO_4)_3$ | $Na_2CO_3 \cdot 10H_2O$ | $FeSO_4 \cdot 7H_2O$ |

## Percentage Composition

From the formula of a compound, we can work out the percentage by mass of each element present in the compound.

**Example 1** Calculate the percentage of silicon and oxygen in silicon(IV) oxide (silica).

**Method:** First, work out the relative molecular mass.
The formula is $SiO_2$

1 atom of silicon, relative atomic mass 28 = 28

2 atoms of oxygen, relative atomic mass 16 = 32

Total = Relative molecular mass = 60

$$\text{Percentage of silicon} = \frac{28}{60} \times 100 = \frac{7}{15} \times 100 = \frac{7 \times 20}{3} = 47\%$$

$$\text{Percentage of oxygen} = \frac{32}{60} \times 100 = \frac{8}{15} \times 100 = \frac{8 \times 20}{3} = 53\%$$

Since every formula unit of silicon(IV) oxide is 47% silicon, and all formula units are identical, bulk samples of pure silicon(IV) oxide all contain 47% silicon. This is true whether you are talking about silicon(IV) oxide found as quartz, or amethyst or crystoballite or sand.

In general,
Percentage of element $A$ =

$$\frac{\text{Relative atomic mass of } A \times \text{No. of atoms of } A \text{ in formula}}{\text{Relative molecular mass of compound}} \times 100$$

**Example 2** Find the percentage by mass of magnesium, oxygen and sulphur in magnesium sulphate.

**Method:** First calculate the relative molecular mass.
The formula is $MgSO_4$

1 atom of magnesium, relative atomic mass 24 = 24

1 atom of sulphur, relative atomic mass 32 = 32

4 atoms of oxygen, relative atomic mass 16 = 64

Total = Relative molecular mass, $M_r$ = 120

$$\text{Percentage of magnesium} = \frac{A_r(\text{Mg}) \times \text{No. of Mg atoms}}{M_r(\text{MgSO}_4)} \times 100$$

$$= \frac{24}{120} \times 100 = \frac{2}{10} \times 100 = 20\%$$

$$\text{Percentage of sulphur} = \frac{A_r(\text{S}) \times \text{No. of S atoms}}{M_r(\text{MgSO}_4)} \times 100$$

$$= \frac{32}{120} \times 100 = \frac{8}{30} \times 100$$

$$= 27\%$$

$$\text{Percentage of oxygen} = \frac{A_r(\text{O}) \times \text{No. of O atoms}}{M_r(\text{MgSO}_4)} \times 100$$

$$= \frac{16 \times 4}{120} \times 100 = \frac{16}{30} \times 100$$

$$= 53\%$$

**Answer:** Magnesium 20%; Sulphur 27%; Oxygen 53%. You can check on the calculation by adding up the percentages to see whether they add up to 100. In this case 20 + 27 + 53 = 100.

**Example 3** Calculate the percentage of water in copper sulphate crystals.

**Method:** Find the relative molecular mass.
The formula is $CuSO_4{\cdot}5H_2O$

1 atom of copper, relative atomic mass 64 = 64 (approx.)

1 atom of sulphur, relative atomic mass 32 = 32

4 atoms of oxygen, relative atomic mass 16 = 64

5 molecules of water, 5 × [(2 × 1) + 16] = 5 × 18 = 90

Total = Relative molecular mass = 250

Mass of water = 90

$$\text{Percentage of water} = \frac{\text{Mass of water in formula}}{\text{Relative molecular mass}} \times 100$$

$$= \frac{90}{250} \times 100 = \frac{9}{25} \times 100$$

$$= 9 \times 4 = 36\%$$

**Answer:** The percentage of water in copper sulphate crystals is 36%.

# Problems on Percentage Composition

## Section 1

Calculators are not needed for these problems.

1. Calculate the percentages of magnesium and oxygen in magnesium oxide, using the expression

$$\text{Percentage of element} = \frac{A_r \text{ of element} \times \text{No. of atoms of element}}{M_r \text{ of compound}} \times 100$$

2. Calculate the percentages by mass of calcium, carbon and oxygen in calcium carbonate.

3. Find the percentages by mass of potassium, hydrogen, carbon and oxygen in potassium hydrogencarbonate, $KHCO_3$.

4. Find the percentages by mass of
   (a) nitrogen and oxygen in nitrogen monoxide, $NO$
   (b) hydrogen and fluorine in hydrogen fluoride, $HF$
   (c) beryllium and oxygen in beryllium oxide, $BeO$
   (d) lithium and oxygen in lithium oxide, $Li_2O$.

5. Calculate the percentages by mass of
   (a) carbon and hydrogen in ethane, $C_2H_6$
   (b) sodium, oxygen and hydrogen in sodium hydroxide, $NaOH$
   (c) sulphur and oxygen in sulphur trioxide, $SO_3$
   (d) carbon and hydrogen in propyne, $C_3H_4$.

6. Calculate the percentages by mass of
   (a) carbon and hydrogen in heptane, $C_7H_{16}$
   (b) magnesium and nitrogen in magnesium nitride, $Mg_3N_2$
   (c) sodium and iodine in sodium iodide, $NaI$
   (d) calcium and bromine in calcium bromide, $CaBr_2$.

# Section 2

These problems can be solved without the use of calculators.

1. Calculate the percentage by mass of
   (a) carbon and hydrogen in pentene, $C_5H_{10}$
   (b) nitrogen, hydrogen and oxygen in ammonium nitrate
   (c) iron, oxygen and hydrogen in iron(II) hydroxide
   (d) carbon, hydrogen and oxygen in ethanedioic acid, $C_2O_4H_2$.

2. Find the percentage by mass of
   (a) iron, sulphur and oxygen in iron(III) sulphate
   (b) water in chromium(III) nitrate-9-water, $Cr(NO_3)_3{\cdot}9H_2O$
   (c) water in sodium sulphide-9-water, $Na_2S{\cdot}9H_2O$
   (d) silicon in silicon(IV) oxide, $SiO_2$.

3. Calculate the percentages of
   (a) carbon, hydrogen and oxygen in propanol, $C_3H_7OH$
   (b) carbon, hydrogen and oxygen in ethanoic acid, $CH_3CO_2H$
   (c) carbon, hydrogen and oxygen in methyl methanoate, $HCO_2CH_3$
   (d) aluminium and sulphur in aluminium sulphide, $Al_2S_3$.

# Section 3

1. Haemoglobin contains 0.33% by mass of iron. There are 2 Fe atoms in 1 molecule of haemoglobin. What is the relative molecular mass of haemoglobin?

2. An adult's bones weigh about 11 kg, and 50% of this mass is calcium phosphate, $Ca_3(PO_4)_2$. What is the mass of phosphorus in the bones of an average adult?

# 3. The Masses of Solids which React Together

## The Mole

Looking at equations tells us a great deal about chemical reactions. For example,

$$Fe(s) + S(s) \rightarrow FeS(s)$$

tells us that iron and sulphur combine to form iron(II) sulphide, and that one atom of iron combines with one atom of sulphur. Chemists are interested in the exact quantities of substances which react together in chemical reactions. For example, in the reaction between iron and sulphur, if you want to measure out just enough iron to combine with, say, 10 g of sulphur, how do you go about it? What you need to do is to count out equal numbers of atoms of iron and sulphur. This sounds a formidable task, and it puzzled a chemist called Avogadro, working in Italy early in the nineteenth century. He managed to solve this problem with a piece of clear thinking which makes the problem look very simple once you have followed his argument.

Avogadro reasoned in this way:

We know from their relative atomic masses that an atom of carbon is 12 times as heavy as an atom of hydrogen. Therefore, we can say:

If 1 atom of carbon is 12 times as heavy as 1 atom of hydrogen,
then 1 dozen C atoms are 12 times as heavy as 1 dozen H atoms,
and 1 hundred C atoms are 12 times as heavy as 1 hundred H atoms,
and 1 million C atoms are 12 times as heavy as 1 million H atoms,

and it follows that when we see a mass of carbon which is 12 times as heavy as a mass of hydrogen, the two masses must contain equal numbers of atoms. If we have 12 g of carbon and 1 g of hydrogen, we know that we have the same number of atoms of carbon and hydrogen. The same argument applies to any element. When we take the relative atomic mass of an element in grams:

| 40 g Calcium | 24 g Magnesium | 32 g Sulphur | 12 g Carbon | 1 g Hydrogen |
|---|---|---|---|---|

all these masses contain the same number of atoms. This number is $6.022 \times 10^{23}$. The amount of an element which contains this number of atoms is called one **mole** of the element. (The symbol for *mole* is **mol.**) The ratio $6.022 \times 10^{23}$/mol is called the **Avogadro constant**. We can count out $6 \times 10^{23}$ atoms of any element by weighing out its relative atomic mass in grams. If we want to react iron and sulphur so that there is an atom of sulphur for every atom of iron, we can count out

$6 \times 10^{23}$ atoms of sulphur by weighing out 32 g of sulphur and we can count out $6 \times 10^{23}$ atoms of iron by weighing out 56 g of iron. Since one atom of iron reacts with one atom of sulphur to form one formula unit of iron(II) sulphide, one mole of iron reacts with one mole of sulphur to form one mole of iron(II) sulphide:

$$Fe(s) + S(s) \rightarrow FeS(s)$$

and 56 g of iron react with 32 g of sulphur to form 88 g of iron(II) sulphide.

Just as one mole of an element is the relative atomic mass in grams, one mole of a compound is the relative molecular mass in grams. If you want to weigh out one mole of sodium hydroxide, you first work out its relative molecular mass.

*The formula is NaOH*
1 atom of Na, relative atomic mass 23 = 23
1 atom of O, relative atomic mass 16 = 16
1 atom of H, relative atomic mass 1 = 1
Total = Relative molecular mass = 40

If you weigh out 40 g of sodium hydroxide, you have one mole of sodium hydroxide. The quantity 40 g/mol is the **molar mass** of sodium hydroxide. The molar mass of a compound is the relative molecular mass in grams per mole. The molar mass of an element is the relative atomic mass in grams per mole. The molar mass of sodium hydroxide is 40 g/mol, and the molar mass of sodium is 23 g/mol.

Remember that most gaseous elements consist of molecules, not atoms. Chlorine exists as $Cl_2$ molecules, oxygen as $O_2$ molecules, hydrogen as $H_2$ molecules, and so on. To work out the mass of a mole of chlorine molecules, you must use the relative molecular mass of $Cl_2$.

Relative atomic mass of chlorine = 35.5
Relative molecular mass of $Cl_2$ = $2 \times 35.5 = 71$
Mass of 1 mole of chlorine, $Cl_2$ = 71 grams.

The noble gases, helium, neon, argon, krypton and xenon, exist as atoms. Since the relative atomic mass of helium is 4, the mass of 1 mole of helium is 4 g.

# Problems on the Mole

REMEMBER:

$$\text{Number of moles of a substance} = \frac{\text{Mass of the substance}}{\text{Mass of one mole of the substance}}$$

Mass of one mole of an element = Relative atomic mass in grams

Mass of one mole of a compound = Relative molecular mass in grams

Relative atomic masses are listed on p. 85.

# Section 1

1. What are the relative atomic masses of sodium, magnesium and lead? What is the mass of 1 mole of:

   (a) sodium (b) magnesium (c) lead?

2. What are the relative atomic masses of barium, chromium and tin? What is the mass of:

   (a) 0.1 mole of barium (b) 0.1 mole of chromium
   (c) 0.1 mole of tin?

3. Use the relative atomic masses of the elements to calculate the mass of:

   (a) 2 moles of iodine molecules (b) 2 moles of silver
   (c) 2 moles of aluminium (d) 2 moles of mercury.

4. Calculate the mass of 0.25 mole of each of these elements:

   (a) silver (b) sulphur (S atoms) (c) magnesium
   (d) calcium (e) neon.

5. Use the relative atomic masses to find the number of moles of the element in:

   (a) 54 g aluminium (b) 2.4 g titanium (c) 42 g iron
   (d) 54 g silver (e) 13 g zinc.

6. What are the relative molecular masses of the following compounds: carbon dioxide, sulphuric acid, hydrogen chloride and sodium hydroxide? State the mass of:

   (a) 1 mole of carbon dioxide (b) 1 mole of sulphuric acid
   (c) 1 mole of hydrogen chloride (d) 1 mole of sodium hydroxide.

7. Use the relative molecular masses of the compounds to calculate the mass of:

   (a) 1 mole of sodium chloride (b) 0.5 mole of potassium hydroxide
   (c) 4 moles of iron(II) chloride (d) 2.5 moles of sodium carbonate
   (e) 0.1 mole of zinc chloride.

8. For each of the following compounds, work out the relative molecular mass, and then state (a) the mass of one mole of the compound, and (b) the mass of 0.25 mole of the compound:
   calcium chloride, copper(II) carbonate, barium hydroxide, sodium nitrate.

# Section 2

1. State the mass of each element in:

   (a) 0.5 mole chromium (b) 1/7 mole iron
   (c) 1/3 mole carbon (d) 1/4 mole magnesium

(e) 1/7 mole nitrogen molecules (f) 1/4 mole oxygen molecules.
Remember that nitrogen and oxygen exist as diatomic molecules, $N_2$ and $O_2$.

2. Calculate the number of moles of each element in:
   (a) 46 g sodium (b) 130 g zinc (c) 10 g calcium
   (d) 2.4 g magnesium (e) 13 g chromium.

3. Calculate the mass of:
   (a) 1 mole of sodium atoms (b) $\frac{1}{2}$ mole of nitrogen atoms
   (c) $\frac{1}{2}$ mole of nitrogen molecules (d) $\frac{1}{4}$ mole of sulphur atoms
   (e) 0.2 mole of bromine atoms (f) 0.2 mole of bromine molecules.

4. Calculate the number of moles of atoms in:
   (a) 23 g sodium (b) 64 g sulphur (c) 9 g aluminium
   (d) 120 g calcium (e) 12 g magnesium (f) 7 g iron.

5. Find the mass of each element in:
   (a) 10 moles of lead (b) 1/6 mole copper
   (c) 0.1 mole iodine molecules (d) 10 moles hydrogen molecules
   (e) 0.25 mole calcium (f) 0.25 mole bromine molecules
   (g) $\frac{3}{4}$ mole iron (h) 0.20 mole zinc
   (i) $\frac{1}{2}$ mole chlorine molecules (j) 0.1 mole neon.

6. State the amount of substance in moles in:
   (a) 58.5 g sodium chloride (b) 26.5 g anhydrous sodium carbonate
   (c) 50.0 g calcium carbonate (d) 15.9 g copper(II) oxide
   (e) 8.0 g sodium hydroxide (f) 303 g potassium nitrate
   (g) 9.8 g sulphuric acid (h) 499 g copper(II) sulphate-5-water.

7. Given Avogadro's constant is $6 \times 10^{23}$/mol, calculate the number of atoms in:
   (a) 35.5 g of chlorine (b) 27 g of aluminium
   (c) 3.1 g of phosphorus (d) 336 g of iron
   (e) 48 g of magnesium (f) 1.6 g of oxygen
   (g) 0.4 g of oxygen (h) 216 g of silver.

8. How many grams of zinc contain:
   (a) $6 \times 10^{23}$ atoms (b) $6 \times 10^{20}$ atoms?

9. How many grams of aluminium contain:
   (a) $2 \times 10^{23}$ atoms (b) $6 \times 10^{20}$ atoms?

10. What mass of carbon contains:
    (a) $6 \times 10^{23}$ atoms (b) $2 \times 10^{21}$ atoms?

11. Write down:
    (a) the mass of calcium which has the same number of atoms as 12 g of magnesium
    (b) the mass of silver which has the same number of atoms as 3 g of aluminium
    (c) the mass of zinc with the same number of atoms as 1 g of helium
    (d) the mass of sodium which has 5 times the number of atoms in 39 g of potassium.

## Section 3

Use Avogadro constant = $6 \times 10^{23}$/mol

1. Imagine a hardware store is having a sale. The knock-down price of titanium is one billion ($10^9$) atoms for 1 p. How much would you have to pay for 1 milligram ($1 \times 10^{-3}$ g) of titanium?
2. Ethanol, $C_2H_6O$, is the alcohol in alcoholic drinks. If you have 9.2 g of ethanol, how many moles do you have of
   (a) ethanol molecules
   (b) carbon atoms
   (c) hydrogen atoms
   (d) oxygen atoms?
3. A car releases about 5 g of nitrogen oxide, NO, into the air for each mile driven. How many molecules of NO are emitted per mile?
4. How many moles of $H_2O$ are there in 1.00 litre of water?
5. How many moles of $Fe_2O_3$ are there in 1.00 kg of rust?
6. What is the mass of one molecule of water?
7. What is the amount in moles of sucrose, $C_{12}H_{22}O_{11}$, in a one kilogram bag of sugar?

# Calculation of Mass of Reactant or Mass of Product

We can use the idea of the mole to work out what mass of product we shall obtain by starting with a known mass of reactant.

**Example 1** What mass of magnesium oxide is obtained from the complete combustion of 12 g of magnesium?

**Method:** First, we write the equation for the reaction:

$$2Mg(s) + O_2(g) \rightarrow 2MgO(s)$$

The equation tells us that:

2 atoms of magnesium form 2 'formula units' of magnesium oxide

therefore 1 atom of magnesium forms 1 'formula unit' of magnesium oxide

and 1 mole of magnesium forms 1 mole of magnesium oxide.

Since $A_r(Mg) = 24$, 1 mole of Mg weighs 24 g.

Since $M_r(MgO) = (24 + 16) = 40$, 1 mole of MgO weighs 40 g.

Therefore, 24 g of magnesium form 40 g of magnesium oxide.

Therefore, 12 g of magnesium form 20 g of magnesium oxide.

**Answer:** 20 g of magnesium oxide are formed by the complete combustion of 12 g of magnesium.

**Example 2** What mass of zinc sulphate can be obtained from the reaction of 10.0 g of zinc with an excess of dilute sulphuric acid?

**Method:** First, we write the equation:

$$Zn(s) + H_2SO_4(aq) \rightarrow H_2(g) + ZnSO_4(aq)$$

Therefore 1 mole of Zn + 1 mole of $H_2SO_4$ → 1 mole of $H_2$ molecules + 1 mole of $ZnSO_4$

Since $A_r(Zn) = 65$, 1 mole of Zn weighs 65 g.

(We will leave out the information about sulphuric acid and hydrogen because this question does not ask for it.)

Calculate the relative molecular mass of zinc sulphate:

1 atom of zinc ($A_r$ 65) = 65

1 atom of sulphur ($A_r$ 32) = 32

4 atoms of oxygen ($A_r$ 16) = 64

Total = $M_r$ of $ZnSO_4$ = 161

1 mole of $ZnSO_4$ weighs 161 g

then 65 g of zinc forms 161 g of zinc sulphate

therefore 1 g of zinc forms $\frac{161}{65}$ g of zinc sulphate

and 10.0 g of zinc forms $\frac{161}{65} \times 10.0$ g of zinc sulphate

= 24.8 g of zinc sulphate

**Answer:** 24.8 g of zinc sulphate can be obtained from 10.0 g of zinc.

The calculation in this problem is a ratio type of calculation. It is tackled in the same way as you tackle such problems in your mathematics lessons, by the unitary method. For example, take the question:

If 3 packets of crisps cost 39 p, what is the cost of 7 packets of crisps?
You work out: If 3 packets of crisps cost 39 p

1 packet of crisps costs $\frac{39}{3}$ p

and 7 packets of crisps cost $\frac{39}{3} \times 7 = 91$ p

**Example 3** Calculate the mass of carbon dioxide produced by heating 15.0 g of limestone.

**Method:** The equation

$$CaCO_3(s) \rightarrow CaO(s) + CO_2(g)$$

tells us that 1 mole of calcium carbonate forms 1 mole of carbon dioxide.

Using $A_r(Ca) = 40$, $A_r(C) = 12$, $A_r(O) = 16$,

$M_r$ of calcium carbonate = $[40 + 12 + (3 \times 16)] = 100$

$M_r$ of carbon dioxide = $[12 + (2 \times 16)] = 44$

1 mole of $CaCO_3$ weighs 100 g, and 1 mole of $CO_2$ weighs 44 g; therefore

100 g of calcium carbonate form 44 g of carbon dioxide

1 g of calcium carbonate forms $\frac{44}{100}$ g of carbon dioxide

15.0 g of calcium carbonate form $\frac{44}{100} \times 15.0$ g of carbon dioxide

= 6.6 g of carbon dioxide

**Answer:** 6.6 g of carbon dioxide will be produced.

**Example 4** If 4.20 g of sodium hydrogencarbonate are heated, what mass of anhydrous sodium carbonate will be formed?

**Method:** First, write the equation:

$$2NaHCO_3(s) \rightarrow Na_2CO_3(s) + CO_2(g) + H_2O(g)$$

This shows that 2 moles of sodium hydrogencarbonate give 1 mole of sodium carbonate.

$M_r$ of $NaHCO_3 = 23 + 1 + 12 + (3 \times 16) = 84$

$M_r$ of $Na_2CO_3 = (2 \times 23) + 12 + (3 \times 16) = 106$

$2 \times 84$ g sodium hydrogencarbonate give 106 g sodium carbonate.

If 168 g sodium hydrogencarbonate give 106 g sodium carbonate

1 g sodium hydrogencarbonate gives $\frac{106}{168}$ g sodium carbonate

4.20 g sodium hydrogencarbonate give $\frac{4.20 \times 106}{168}$ g sodium carbon

$= \frac{0.6 \times 106}{24} = \frac{0.1 \times 106}{4}$

= 2.65 g

**Answer:** 2.65 g of sodium carbonate will be formed.

# Using the Masses of the Reactants to work out the Equation for a Reaction

The equation for a reaction can be used to enable you to calculate the masses of chemicals taking part in the reaction. The converse is also true. If you know the mass of each substance taking part in a reaction, you can calculate the number of moles of each substance taking part in the reaction, and this will tell you the equation.

**Example 1** Iron burns in chlorine to form iron chloride. An experiment showed that 5.60 g of iron combined with 10.65 g of chlorine. Deduce the equation for the reaction.

**Method:**

5.60 g of iron combine with 10.65 g of chlorine

Relative atomic masses are: Fe = 56, Cl = 35.5

Number of moles of iron = 5.60/56 = 0.10

Number of moles of chlorine atoms = 10.65/35.5 = 0.30

The equation must be:

$$Fe + 3Cl \rightarrow$$

Since chlorine exists as $Cl_2$ molecules, we must multiply by 2:

$$2Fe + 3Cl_2 \rightarrow$$

To balance the equation, the right-hand side must read $2FeCl_3$. Therefore,

**Answer:** $2Fe(s) + 3Cl_2(g) \rightarrow 2FeCl_3(s)$

**Example 2** 17.0 g of sodium nitrate react with 19.6 g of sulphuric acid to give 12.6 g of nitric acid. Deduce the equation for the reaction.

**Method:**

Relative molecular masses are: $NaNO_3$ = 85, $H_2SO_4$ = 98, $HNO_3$ = 63

Number of moles of $NaNO_3$ = 17.0/85 = 0.2

Number of moles $H_2SO_4$ = 19.6/98 = 0.2

Number of moles of $HNO_3$ = 12.6/63 = 0.2

0.2 mol $NaNO_3$ reacts with 0.2 mol $H_2SO_4$ to form 0.2 mol of $HNO_3$

1 mol $NaNO_3$ reacts with 1 mol $H_2SO_4$ to form 1 mol of $HNO_3$

$$NaNO_3(a) + H_2SO_4(l) \rightarrow HNO_3(l)$$

The equation must be balanced by inserting $NaHSO_4$ on the right-hand side:

**Answer:** $NaNO_3(s) + H_2SO_4(l) \rightarrow HNO_3(l) + NaHSO_4(s)$

# Problems on Reacting Masses of Solids

## Section 1

These problems can be solved without calculators.

1. What mass of magnesium oxide is formed by the complete combustion of 24 g of magnesium? Write the equation. Put in the relative atomic masses. Out comes the answer.

2. What mass of magnesium oxide is formed by the complete combustion of 6 g of magnesium? Write the equation. Put in the relative atomic masses. Do a ratio type of calculation.

3. What mass of carbon dioxide is formed by the complete combustion of 12 g of carbon? Write the equation. Put in the relative atomic masses.

4. What mass of carbon dioxide is formed by the complete combustion of 4 g of carbon? Write the equation. Put in the relative atomic masses. Do a ratio type of calculation.

5. Calculate the mass of sulphur that must be burned to produce 64 g of sulphur dioxide. First, write the equation. Then put in the relative atomic masses.

6. Calculate the mass of sulphur that must be burned to give 8 g of sulphur dioxide.

7. Calculate the mass of sulphur that must be burned to give 100 g of sulphur dioxide.

8. In the reduction of copper(II) oxide by hydrogen, according to the equation

$$CuO(s) + H_2(g) \rightarrow Cu(s) + H_2O(g)$$

what mass of copper can be obtained from:

(a) 79.5 g of copper(II) oxide (b) 15.9 g of copper(II) oxide?

Look up the relative masses, and use them in the equation.

9. What mass of carbon dioxide can be produced by heating 10 g of calcium carbonate? Write the equation. Put in the relative atomic masses, and do a ratio type of calculation.

10. What mass of hydrogen can be made by reacting 12 g of magnesium with an excess of dilute sulphuric acid? First, the equation; second, the relative atomic masses; third, a ratio type of calculation.

11. What mass of carbon can be completely burned in 32 g of oxygen? There are three steps to remember.

12. What mass of iron must be heated with excess sulphur to produce 4.4 g of iron(II) sulphide? You should remember the three steps by now.

## Section 2

These problems can be solved without calculators.

1. Calculate what mass of carbon you would need to reduce 15.9 g copper(II) oxide to copper by the reaction

$$CuO(s) + C(s) \rightarrow Cu(s) + CO(g)$$

2. Iron will react with chlorine to form iron(III) chloride:

$$2Fe(s) + 3Cl_2(g) \rightarrow 2FeCl_3(s)$$

Find the mass of iron(III) chloride that can be obtained from 8 g iron.

3. A mixture of 8 g iron and 4 g sulphur is heated, and the elements react to form iron(II) sulphide, FeS. How much iron will be left over at the end of the reaction?

4. What mass of lead(II) oxide is obtained by heating 33.1 g of lead(II) nitrate in the reaction

$$2Pb(NO_3)_2(s) \rightarrow 2PbO(s) + 4NO_2(g) + O_2(g)$$

5. Calculate the mass of carbon dioxide you can obtain by the action of acid on 15 g calcium carbonate, in the reaction

$$CaCO_3(s) + 2HCl(aq) \rightarrow CO_2(g) + CaCl_2(aq) + H_2O(l)$$

6. Calculate what mass of sodium hydroxide you would need to neutralise a solution containing 7.3 g hydrogen chloride by the reaction

$$NaOH(aq) + HCl(aq) \rightarrow NaCl(aq) + H_2O(l)$$

7. Find how many grams of sodium sulphate are formed when 49 g of sulphuric acid are dissolved and neutralised by sodium hydroxide solution:

$$H_2SO_4(aq) + 2NaOH(aq) \rightarrow Na_2SO_4(aq) + 2H_2O(l)$$

8. What mass of zinc chloride, $ZnCl_2$, is formed when 13 g zinc are completely converted into the chloride?

9. Calculate the mass of potassium chloride formed when a solution containing 8.00 g potassium hydroxide is neutralised with hydrochloric acid

$$KOH(aq) + HCl(aq) \rightarrow KCl(aq) + H_2O(l)$$

10. Calculate how much sodium nitrate you need to give 126 g of nitric acid by the reaction

$$NaNO_3(s) + H_2SO_4(l) \rightarrow HNO_3(l) + NaHSO_4(s)$$

11. Which one of the following contains the same number of atoms as 7 g of iron?
    - **A** 4 g of aluminium
    - **B** 4 g of magnesium
    - **C** 4 g of sulphur
    - **D** 3 g of carbon
    - **E** 4 g of calcium

12. State which of these compounds contains the largest percentage by mass of nitrogen:
    - **A** ammonium chloride, $NH_4Cl$
    - **B** ammonium nitrate, $NH_4NO_3$
    - **C** ammonium sulphate, $(NH_4)_2SO_4$
    - **D** ammonia, $NH_3$
    - **E** urea, $CO(NH_2)_2$

13. Which of the following contains the same number of atoms as 10 g of calcium?
    - **A** 6 g of sodium
    - **B** 13 g of chromium
    - **C** 8 g of magnesium
    - **D** 26 g of silver
    - **E** 7 g of aluminium

# Section 3

1. A sulphuric acid plant uses 2500 tonne of sulphur dioxide each day. What mass of sulphur must be burned to produce this quantity of sulphur dioxide?

2. An antacid tablet contains 0.1 g of magnesium hydrogencarbonate, $Mg(HCO_3)_2$. What mass of stomach acid, HCl, will it neutralise?

3. Wine is made by the fermentation of the sugar in grapes:

   sucrose (sugar) → ethanol (in wine) + carbon dioxide

   $$C_6H_{12}O_6(aq) \rightarrow 2C_2H_6O(aq) + 2CO_2(g)$$

   What mass of ethanol can be obtained from 6.00 kg of sucrose?

4. Aspirin, $C_9H_8O_4$, is made by the reaction:

   salicylic acid + ethanoic anhydride → aspirin + ethanoic acid

   $$C_7H_6O_3 + C_4H_6O_3 \rightarrow C_9H_8O_4 + C_2H_4O_2$$

   How many grams of salicylic acid, $C_7H_6O_3$, are needed to make one aspirin tablet, which contains 0.33 g of aspirin?

5. Natural gas (methane, $CH_4$) burns according to the *unbalanced* equation:

   $$CH_4(g) + O_2(g) \rightarrow CO_2(g) + H_2O(l)$$

(a) Balance the equation.

(b) Say what mass of water will be formed when 1.00 kg of natural gas is burned completely.

6. Aluminium sulphate is used to treat sewage. It can be made by the reaction:

aluminium hydroxide + sulphuric acid → aluminium sulphate + water

(a) Balance this *unbalanced* equation for the reaction:

$$Al(OH)_3(s) + H_2SO_4(aq) \rightarrow Al_2(SO_4)_3(aq) + H_2O(l)$$

(b) Say what masses of (i) aluminium hydroxide and (ii) sulphuric acid are needed to make 1.00 kg of aluminium sulphate.

7. The element phosphorus is obtained from the ore calcium phosphate by means of the reaction:

calcium phosphate + silicon(IV) oxide + carbon (coke) → calcium silicate + carbon monoxide + phosphorus

$$Ca_3(PO_4)_2(s) + 3SiO_2(s) + 5C(s) \rightarrow 3CaSiO_3(s) + 5CO(g) + 2P(s)$$

What mass of calcium phosphate must be used to yield 5.00 kg of phosphorus?

8. Washing soda, $Na_2CO_3{\cdot}10H_2O$, loses some of its water of crystallisation if it is not kept in an air-tight container to form $Na_2CO_3{\cdot}H_2O$.

A grocer buys a 10 kg bag of washing soda at 30 p/kg. While it is standing in his store room, the bag punctures, and the crystals turn into a powder, $Na_2CO_3{\cdot}H_2O$. The grocer sells this powder at 50 p/kg. Does he make a profit or a loss?

9. When you take a warm bath, the power station has to burn about 1.2 kg of coal to provide enough electricity to heat the water.

(a) If the coal contains 3% sulphur, what mass of sulphur dioxide does the power station emit as a result?

(b) Multiply your answer by the number of warm baths you take in a year.

(c) This is only a part of your contribution to air pollution. What can be done to reduce this source of pollution – apart from taking cold baths?

10. A layer of ozone in the upper atmosphere protects us from receiving too much ultraviolet radiation from the sun. Ozone, $O_3$, is formed in the upper atmosphere from oxygen, $O_2$:

$$O_2(g) \rightarrow O_3(g)$$

(a) Balance the equation.

(b) Say what mass of ozone is formed from 64 g of oxygen.

11. Nitrogen monoxide, NO, is a pollutant gas which comes out of vehicle exhausts. One technique for reducing the quantity of nitrogen monoxide in vehicle exhausts is to inject a stream of ammonia, $NH_3$,

into the exhaust. Nitrogen monoxide is converted into the harmless products, nitrogen and water:

$$4NH_3(g) + 6NO(g) \rightarrow 5N_2(g) + 6H_2O(l)$$

An average vehicle emits 5 g of nitrogen monoxide per mile. Assuming a mileage of 10 000 miles a year, what mass of ammonia would be needed to clean up the exhaust?

12. The ore chalcopyrite, $CuFeS_2$, is a source of copper. A copper smelter can produce 500 tonne of copper a day by the reaction:

$$2CuFeS_2(s) + 5O_2(g) \rightarrow 2Cu(s) + 2FeO(s) + 4SO_2(g)$$

The sulphur dioxide produced is released into the atmosphere and eventually converted into sulphuric acid by the reactions:

$$2SO_2(g) + O_2(g) \rightarrow 2SO_3(g)$$

$$SO_3(g) + H_2O(l) \rightarrow H_2SO_4(l)$$

(a) If a smelter produces 500 tonnes of copper in a day, what is the output of sulphur dioxide?

(b) What mass of sulphuric acid is produced in a day in the absence of any clean-up system?

**Use** $A_r(Cu) = 64$

13. Some industrial plants, for example aluminium smelters, emit fluorides. In the past, there have been cases of fluoride pollution affecting the teeth and joints of cattle. The Union Carbide Corporation has invented a process for removing fluorides from waste gases. It involves the reaction:

$$\text{fluoride ion } (F^-) + \text{charcoal (C)} \xrightarrow{\text{high temperature}} \text{carbon tetrafluoride } (CF_4)$$

The product, $CF_4$, is harmless. The firm claims that 1 kg of charcoal will remove 6.3 kg of fluoride ion. Do you believe this claim? Explain your answer.

14. A factory makes a detergent of formula $C_{12}H_{25}SO_4Na$ from lauryl alcohol, $C_{12}H_{26}O$. To manufacture 11 tonnes of detergent daily, what mass of lauryl alcohol is needed?

15. A mass of 0.65 g of zinc powder was added to a beaker containing silver nitrate solution. When all the zinc had reacted, 2.16 g of silver were obtained.

(a) Calculate the number of moles of zinc used.

(b) Calculate the number of moles of silver formed.

(c) Calculate the number of moles of silver produced by 1 mole of zinc.

(d) Write a balanced ionic equation for the reaction.

16. The element X has a relative atomic mass of 35.5. It reacts with a solution of the sodium salt of Y according to the equation

$$X_2 + 2NaY \rightarrow Y_2 + 2NaX$$

If 14.2 g of $X_2$ displace 50.8 g of $Y_2$, what is the relative atomic mass of Y?

17. Silicon reacts with chlorine to form silicon tetrachloride, $SiCl_4$.
    (a) If 3.40 g of $SiCl_4$ is formed, what mass of silicon has reacted?
    (b) If 3.40 g of $SiCl_4$ react with water, how many moles of hydrogen chloride are formed?

$$SiCl_4(l) + 2H_2O(l) \rightarrow SiO_2(s) + 4HCl(g)$$

18. TNT is an explosive. The name stands for trinitrotoluene. The compound is made by the reaction

toluene + nitric acid → TNT + water

$$C_7H_8(l) + 3HNO_3(l) \rightarrow C_7H_5N_3O_6(s) + 3H_2O(l)$$

Calculate the masses of (a) toluene and (b) nitric acid that must be used to make 10.00 tonnes of TNT. (1 tonne = 1000 kg)

19. A large power plant produces about 500 tonnes of sulphur dioxide in a day. One way of removing this pollutant from the waste gases is to inject limestone. This converts sulphur dioxide into calcium sulphate.

(i) limestone + sulphur dioxide + oxygen → calcium sulphate + carbon dioxide

$$2CaCO_3(s) + 2SO_2(g) + O_2(g) \rightarrow 2CaSO_4(s) + 2CO_2(g)$$

Another method of removing sulphur dioxide is to 'scrub' the waste gases with ammonia. The product is ammonium sulphate.

(ii) ammonia + sulphur dioxide + oxygen + water → ammonium sulphate

$$4NH_3(g) + 2SO_2(g) + O_2(g) + 2H_2O(l) \rightarrow 2(NH_4)_2SO_4(aq)$$

(a) Calculate the mass of calcium sulphate produced in a day by method (i) and the mass of ammonium sulphate produced in a day by method (ii).

(b) Say whether limestone and ammonia are found naturally occurring or are manufactured.

(c) Which do you think is the more valuable by-product, calcium sulphate or ammonium sulphate? Explain your answer.

# 4. Finding Formulae

## Empirical Formulae

From the formula, we can find the mass of each element present in a certain mass of the compound. The reverse is also true: from the mass of each element present in a sample of the compound, we can find the formula of the compound. The method uses the ratio type of calculation. For example:

If the relative atomic mass of magnesium is 24, how many moles of magnesium are there in 6 g of magnesium?

If 24 g of magnesium are 1 mole,
then 6 g of magnesium are 6/24 mole = $\frac{1}{4}$ mole.

In general,

$$\text{Number of moles of substance} = \frac{\text{Mass of substance}}{\text{Mass of one mole of the substance}}$$

Mass of one mole of an element = Relative atomic mass in grams

Mass of one mole of a compound = Relative molecular mass in grams

**Example 1** Given that 127 g of copper combine with 32 g of oxygen, what is the formula of copper oxide?

| *Elements* | *Copper* | | *Oxygen* |
|---|---|---|---|
| Symbols | Cu | | O |
| Masses | 127 g | | 32 g |
| Relative atomic masses | 63.5 | | 16 |
| Number of moles | $\frac{127}{63.5}$ = 2 | | $\frac{32}{16}$ = 2 |
| Divide through by 2 | = 1 mole | to | 1 mole |
| Number of atoms | = 1 atom | to | 1 atom |
| Formula | | CuO | |

We divide through by two to obtain the simplest formula for copper oxide which will fit the data. **The simplest formula which represents the composition of a compound is called the empirical formula.**

**Example 2** Given that 0.96 g of magnesium combines with 2.84 g of chlorine, what is the empirical formula for magnesium chloride?

| *Elements* | *Magnesium* | | *Chlorine* |
|---|---|---|---|
| Symbols | Mg | | Cl |
| Masses | 0.96 g | | 2.84 g |
| Relative atomic masses | 24 | | 35.5 |
| Number of moles | $\frac{0.96}{24}$ | | $\frac{2.84}{35.5}$ |
| | = 0.04 | | = 0.08 |
| Divide through by the smaller number | = 1 mole | to | 2 moles |
| Number of atoms | = 1 atom | to | 2 atoms |
| Empirical formula | | $MgCl_2$ | |

**Example 3** If the percentage of water in magnesium sulphate crystals is 51.2%, what is $n$ in the formula $MgSO_4{\cdot}nH_2O$?

*Note* that, when we say that the percentage of water in the crystals is 51.2%, we mean that 100 g of crystals contain 51.2 g of water. The difference, 48.8 g is the mass of magnesium sulphate.

| *Compounds* | *Magnesium sulphate* | | *Water* |
|---|---|---|---|
| Formulae | $MgSO_4$ | | $H_2O$ |
| Masses | 48.8 g | | 51.2 g |
| Relative molecular masses | 120 | | 18 |
| Number of moles | $\frac{48.8}{120}$ | | $\frac{51.2}{18}$ |
| | = 0.406 | | = 2.85 |
| Divide through by the smaller number | = 1 mole | to | 7 moles |
| Empirical formula | | $MgSO_4{\cdot}7H_2O$ | |

**Example 4** When 127 g of copper combine with oxygen, 143 g of an oxide are formed. What is the empirical formula of the oxide?

**Method:** You will notice here that the mass of oxygen is not given to you. You obtain it by subtraction.

Mass of copper = 127 g

Mass of oxide = 143 g

Mass of oxygen = 143 − 128 = 16 g.

Now you can carry on as before:

| *Elements* | *Copper* | *Oxygen* |
|---|---|---|
| Symbols | Cu | O |
| Masses | 127 g | 16 g |

| | | | |
|---|---|---|---|
| Relative atomic masses | 63.5 | | 16 |
| Number of moles | $\frac{127}{63.5}$ | | $\frac{16}{16}$ |
| | = 2 | | = 1 |
| Number of atoms | = 2 | to | 1 |

**Answer:** Empirical formula $Cu_2O$

# How to find the Molecular Formula from the Empirical Formula

The molecular formula of a compound can be found from the empirical formula if the relative molecular mass is known.

**Example 1** Analysis shows the empirical formula of a compound to be $CH_2O$. Its relative molecular mass is 60. What is its molecular formula?

**Method:**

Relative molecular mass = 60

Relative empirical formula mass = (12 + 2 + 16) = 30

The relative molecular mass is double the relative empirical formula mass.

The molecular formula is double the empirical formula.

**Answer:** The molecular formula is $C_2H_4O_2$.

**Example 2** What is the molecular formula of the compound, A, which has an empirical formula $C_2H_6O$ and a relative molecular mass of 46?

**Method:**

Relative molecular mass of A = 46

Relative empirical formula mass = $(2 \times 12) + 6 + 16 = 46$

**Answer:** The empirical formula gives the correct relative molecular mass; therefore the molecular formula is the same as the empirical formula, $C_2H_6O$.

# Problems on Formulae

## Section 1

These examples can be done without calculators.

1. 2.3 g of sodium combine with 0.80 g of oxygen.
   How many moles of sodium does this mass represent?
   How many moles of oxygen atoms are involved?
   How many moles of sodium combine with one mole of oxygen?
   What is the formula of sodium oxide?

2. 0.72 g of magnesium combine with 0.28 g of nitrogen.
   How many moles of magnesium does this represent?
   How many moles of nitrogen atoms combine?
   How many moles of magnesium combine with one mole of nitrogen atoms?
   What is the formula of magnesium nitride?

3. 1.68 g of iron combine with 0.64 g of oxygen.
   How many moles of iron does this mass represent?
   How many moles of oxygen atoms combine?
   How many moles of iron combine with one mole of oxygen atoms?
   What is the formula of this oxide of iron?

4. A compound contains 55.5% by mass of mercury; 44.5% by mass of bromine.
   How many grams of mercury are there in 100 g of the compound?
   How many grams of bromine are there in 100 g of the compound?
   How many moles of mercury are there in 100 g of the compound?
   How many moles of bromine atoms are there in 100 g of the compound?
   Work out the ratio of moles of bromine atoms to moles of mercury atoms.
   What is the formula of the compound?

5. Calculate the empirical formula of the compound formed when 2.70 g of aluminium form 5.10 g of its oxide.
   What is the mass of aluminium?
   What is the mass of oxygen (not oxide)?
   How many moles of aluminium combine?
   How many moles of oxygen atoms combine?
   What is the ratio of moles of aluminium to moles of oxygen atoms?
   What is the formula of aluminium oxide?

6. Barium chloride forms a hydrate which contains 85.25% barium chloride and 14.75% water of crystallisation. What is the formula of this hydrate?
   What is the mass of barium chloride in 100 g of the hydrate?
   What is the mass of water in 100 g of the hydrate?
   What is the relative molecular mass of barium chloride?
   What is the relative molecular mass of water?
   How many moles of barium chloride are present in 100 g of the hydrate?
   How many moles of water are present in 100 g of the hydrate?
   What is the ratio of moles of barium chloride to moles of water?
   What is the formula of barium chloride hydrate?

7. Calculate the empirical formula of the compound formed when 414 g of lead form 478 g of a lead oxide.
   What mass of lead is present?
   How many moles of lead are present?

What mass of oxygen (not oxide) is present?
How many moles of oxygen atoms are present?
What is the formula of this oxide of lead?

8. Calculate the empirical formulae of the compounds which analyse as
(a) 84% carbon, 16% hydrogen
(b) 72% magnesium, 28% nitrogen
(c) 36% aluminium, 64% sulphur
(d) 20% calcium, 80% bromine
(e) 52% chromium, 48% sulphur

# Section 2

No calculators are required.

1. Calculate the empirical formulae of the compounds containing
(a) sulphur 50%, oxygen 50%
(b) sulphur 40%, oxygen 60%
(c) nitrogen 47%, oxygen 53%
(d) nitrogen 30.5%, oxygen 69.5%
(e) carbon 75%, hydrogen 25%
(f) carbon 85.7%, hydrogen 14.3%

2. Calculate the empirical formulae of the following compounds:
(a) 0.62 g of phosphorus combined with 0.48 g of oxygen
(b) 1.4 g of nitrogen combined with 0.30 g of hydrogen
(c) 0.62 g of lead combined with 0.064 g of oxygen
(d) 3.5 g of silicon combined with 4.0 g of oxygen
(e) 1.10 g of manganese combined with 0.64 g of oxygen
(f) 4.2 g of nitrogen combined with 12.0 g of oxygen
(g) 2.6 g of chromium combined with 5.3 g of chlorine

3. Find the molecular formula for each of the following compounds from the empirical formula and the relative molecular mass:

| | Empirical formula | $M_r$ | | Empirical formula | $M_r$ |
|---|---|---|---|---|---|
| *A* | $CF_2$ | 100 | *E* | $CH_2$ | 42 |
| *B* | $C_2H_4O$ | 88 | *F* | $CH_3O$ | 62 |
| *C* | $CH_3$ | 30 | *G* | $CH_2Cl$ | 99 |
| *D* | $CH$ | 78 | *H* | $C_2HNO_2$ | 213 |

4. Calculate the empirical formulae of the compounds formed when
   (a) 0.69 g of sodium forms 0.93 g of an oxide of sodium
   (b) 10.35 g of lead form 11.41 g of an oxide of lead
   (c) 0.035 g of nitrogen forms 0.115 g of an oxide of nitrogen
   (d) 2.54 g of copper form 2.86 g of an oxide
   (e) 11.2 g of iron form 25.4 g of a chloride of iron
   (f) 14.0 g of iron combine with 26.6 g of chlorine

5. Calculate the empirical formulae of the compounds formed when
   (a) 0.24 g of carbon combines with 0.64 g of oxygen
   (b) 20.7 g of lead form 23.9 g of a lead oxide
   (c) 15.9 g of copper combine with 17.7 g of chlorine
   (d) 6 g of magnesium combine with 4 g of oxygen
   (e) 1.8 g of magnesium form 2.5 g of magnesium nitride
   (f) 9 g of aluminium form 89 g of aluminium bromide.

6. Calculate the empirical formulae of these hydrates:
   (a) magnesium sulphate crystals, which contain 48.8% of magnesium sulphate and 51.2% of water
   (b) copper sulphate crystals, which contain 63.9% of copper sulphate and 36.1% of water
   (c) crystals of chromium(III) nitrate, which contain 59.5% of chromium(III) nitrate and 40.5% of water.

7. Calculate the empirical formulae of the compounds with the following compositions:
   (a) 20.0% magnesium, 26.6% sulphur, 53.3% oxygen
   (b) 35.0% nitrogen, 5.0% hydrogen, 60.0% oxygen
   (c) 60.0% carbon, 13.3% hydrogen, 26.7% oxygen
   (d) 40.0% carbon, 6.7% hydrogen, 53.3% oxygen

## Section 3

1. 16.25 g of a chloride of iron were produced from the reaction of 5.6 g of iron with an excess of chlorine. Find the formula of the chloride.

2. Which of the elements (C, H, N, O, Br) is present in the highest proportion by mass in a compound of formula $C_6H_4N_3O_4Br$?

3. Magnesium bromide crystallises as a hydrate, $MgBr_2{\cdot}xH_2O$. When 7.30 g of the hydrate were heated to constant mass, 4.60 g of the anhydrous salt remained. Find the value of $x$.

4. Which one of the following ores contains the highest proportion of aluminium?

| | *Ore* | $M_r$ |
|---|---|---|
| **A** | $Al_2O_3 \cdot H_2O$ | 120 |
| **B** | $Al_2O_3 \cdot 2H_2O$ | 138 |
| **C** | $Na_3AlF_6$ | 210 |
| **D** | $MgAl_2O_4$ | 142 |
| **E** | $KAl(SO_4)_2 \cdot 12H_2O$ | 474 |

5. A metal M forms a chloride of formula $MCl_2$ and relative molecular mass 127. The chloride reacts with sodium hydroxide solution to form a precipitate of the metal hydroxide. What is the relative molecular mass of the hydroxide?

   **A** 56

   **B** 71

   **C** 73

   **D** 90

   **E** 146

6. A porcelain boat was weighed. After a sample of the oxide of a metal M, of $A_r = 119$, was placed in the boat, the boat was reweighed. Then the boat was placed in a reduction tube, and heated while a stream of hydrogen was passed over it. The oxide was reduced to the metal M. The boat was allowed to cool with hydrogen still passing over it, and then reweighed. Then it was reheated in hydrogen, cooled again and reweighed. The following results were obtained.

   Mass of boat = 6.10 g
   Mass of boat + metal oxide = 10.63 g
   Mass of boat + metal (1) = 9.67 g
   Mass of boat + metal (2) = 9.67 g

   (a) Explain why hydrogen was passed over the boat while it was cooling.

   (b) Explain why the boat + metal was reheated.

   (c) Find the empirical formula of the metal oxide.

   (d) Write an equation for the reaction of the oxide with hydrogen.

# 5. The Volumes of Gases which React Together

## Calculating the Reacting Volumes of Gases from the Equation for a Reaction

A surprising feature of reactions between gases was noticed by a French chemist called *Gay-Lussac* in 1808. **Gay-Lussac's Law** *states that when gases combine they do so in volumes which bear a simple ratio to one another and to the volume of the product if it is gaseous, provided all the volumes are measured at the same temperature and pressure.* For example, when hydrogen and chlorine combine, 1 $dm^3$ (or litre) of hydrogen will combine exactly with 1 $dm^3$ of chlorine to form 2 $dm^3$ of hydrogen chloride. When nitrogen and hydrogen combine, a certain volume of nitrogen will combine with three times that volume to form twice its volume of ammonia.

The Italian chemist Avogadro gave an explanation in 1811. His suggestion, known as **Avogadro's Hypothesis**, is that: *Equal volumes of all gases (at the same temperature and pressure) contain the same number of molecules.* It follows from Avogadro's Hypothesis that, whenever we see an equation representing a reaction between gases, we can substitute volumes of gases in the same ratio as numbers of molecules.

Thus, $N_2(g) + 3H_2(g) \rightarrow 2NH_3(g)$

means that since

1 molecule of nitrogen + 3 molecules of hydrogen form 2 molecules of ammonia

then 1 volume of nitrogen + 3 volumes of hydrogen form 2 volumes of ammonia.

For example, 1 $dm^3$ of nitrogen + 3 $dm^3$ of hydrogen form 2 $dm^3$ of ammonia.

Since equal volumes of gases (at the same temperature and pressure) contain the same number of molecules, if you consider the Avogadro constant, $L$ molecules of carbon dioxide, $L$ molecules of hydrogen, $L$ molecules of oxygen and so on, then all the gases will occupy the same volume. The volume occupied by $L$ molecules of gas, which is one mole of each gas, is called the **gas molar volume**. The gas molar volume is 24.0 $dm^3$ at room temperature (20 °C) and 1 atmosphere pressure (r.t.p.). The unit cubic decimetre ($dm^3$) is used for gas molar volume here. Other units are cubic centimetre ($cm^3$) and litre (l).

$$1\ dm^3 = 1000\ cm^3 = 1\ l$$

Calculations on the reacting volumes of gases are very simple as they depend on the fact that

> One mole of gas occupies 24.0 $dm^3$ at room temperature and 1 atmosphere.

**Example 1** What volume of hydrogen (at r.t.p.) is evolved when 0.325 g of zinc reacts with dilute hydrochloric acid?

**Method:** From the equation

$$Zn(s) + 2HCl(aq) \rightarrow H_2(g) + ZnCl_2(aq)$$

we see that 1 mole of zinc gives 1 mole of hydrogen, $H_2$
therefore 65 g of zinc give 24.0 $dm^3$ of hydrogen at r.t.p.

1 g of zinc gives 24.0/65 $dm^3$ of hydrogen at r.t.p.

0.325 g of zinc gives

$$\frac{24.0 \times 0.325}{65} \text{ dm}^3 \text{ of hydrogen at r.t.p.}$$

$$= 0.120 \text{ dm}^3 \text{ or } 120 \text{ cm}^3 \text{ of hydrogen}$$

**Answer:** 120 $cm^3$ of hydrogen.

**Example 2** What volume of oxygen is required for the complete combustion of 125 $cm^3$ of butene? What volume of carbon dioxide is formed? (All volumes are quoted at the same temperature and pressure.)

**Method:** The equation

$$C_4H_8(g) + 6O_2(g) \rightarrow 4CO_2(g) + 4H_2O(g)$$

shows that

1 mole of butene + 6 moles of oxygen form 4 moles of carbon dioxide;
1 volume of butene + 6 volumes of oxygen form 4 volumes of carbon dioxide;
and 125 $cm^3$ of butene + 750 $cm^3$ of oxygen form 500 $cm^3$ of carbon dioxide.

**Answer:** Volume of oxygen = 750 $cm^3$. Volume of carbon dioxide = 500 $cm^3$.

## Using the Reacting Volumes of Gases to work out the Equation for a Reaction

From the volumes of gases taking part in a reaction can be worked out the number of moles of each gas taking part in a reaction. From this the equation follows.

**Example 1** 100 $cm^3$ of propane burn in 500 $cm^3$ of oxygen to form 300 $cm^3$ of carbon dioxide and 400 $cm^3$ of steam. What is the equation for the reaction?

**Method:** Since

100 $cm^3$ propane + 500 $cm^3$ oxygen → 300 $cm^3$ $CO_2$ + 400 $cm^3$ steam,

1 molecule of propane + 5 molecules of oxygen → 3 molecules of $CO_2$ + 4 molecules of steam

and the equation is, therefore,

**Answer:** $C_3H_8(g) + 5O_2(g) \rightarrow 3CO_2(g) + 4H_2O(g)$

**Example 2** 200 cm$^3$ of ammonia burn in 250 cm$^3$ of oxygen to form 200 cm$^3$ of nitrogen monoxide and 300 cm$^3$ of steam. What is the equation for this reaction?

**Method:** Since

200 cm$^3$ ammonia + 250 cm$^3$ oxygen $\rightarrow$ 200 cm$^3$ NO + 300 cm$^3$ steam,

2 volumes of ammonia + 2.5 volumes of oxygen $\rightarrow$ 2 volumes of nitrogen monoxide + 3 volumes of steam

2 molecules of $NH_3$ + 2.5 molecules of $O_2$ $\rightarrow$ 2 molecules of NO + 3 molecules of $H_2O$

To give an equation with whole numbers of molecules, we multiply through by two, and write

4 molecules of $NH_3$ + 5 molecules of $O_2$ $\rightarrow$ 4 molecules of NO + 6 molecules of $H_2O$

**Answer:** $4NH_3(g) + 5O_2(g) \rightarrow 4NO(g) + 6H_2O(g)$

# Problems on Reacting Volumes of Gases

## Section 1

1. The complete combustion of carbon in oxygen yields carbon dioxide. Calculate the volume of carbon dioxide (at r.t.p.) that would be formed by the combustion of 12 g of carbon.
   First, write the equation for the combustion of carbon.
   How many moles of carbon dioxide are formed from 1 mole of carbon?
   What volume of carbon dioxide is formed from 1 mole of carbon?
   How many moles of carbon are there in 12 g of carbon?
   What volume of carbon dioxide will this number of moles of carbon produce?

2. What volume of hydrogen (at r.t.p.) is formed when 12 g of magnesium react with an excess of acid?
   First, write the equation for the reaction of magnesium with any dilute acid.
   How many moles of hydrogen are produced from 1 mole of magnesium?

What is the volume of this number of moles of hydrogen?
How many moles of magnesium are there in 12 g of the metal?
What is the volume of hydrogen produced from this number of moles of magnesium?

3. What volume of hydrogen (at r.t.p.) is produced when 6.5 g of zinc react with an excess of acid?
First, write the equation for the reaction of zinc with a dilute acid.
How many moles of hydrogen are formed from 1 mole of zinc?
What is the volume of this number of moles of hydrogen?
How many moles of zinc are present in 6.5 g of zinc?
What is the volume of hydrogen formed from this number of moles of zinc?

4. What is the volume of carbon dioxide (at r.t.p.) obtained by heating 10.0 g of calcium carbonate?
First, write the equation for the thermal decomposition of calcium carbonate.
How many moles of carbon dioxide are formed from 1 mole of calcium carbonate?
What is the volume of this number of moles of carbon dioxide?
How many moles of calcium carbonate are there in 10 g of this solid?
What volume of carbon dioxide is formed from this number of moles?

5. Calculate the volume of oxygen (at r.t.p.) needed for the complete combustion of 125 $cm^3$ of methane. What volume of carbon dioxide is formed?
First, write the equation for the combustion of methane, $CH_4$.
How many moles of oxygen react with 1 mole of methane?
How many volumes of oxygen react with 1 volume of methane?
What volume of oxygen reacts with 125 $cm^3$ of methane?
How many moles of carbon dioxide are formed by combustion of 1 mole of methane?
How many volumes of carbon dioxide are formed from 1 volume of methane?
What volume of carbon dioxide is formed from 125 $cm^3$ of methane?

6. After an electric spark is passed through a mixture of 250 $cm^3$ of hydrogen and 250 $cm^3$ of oxygen, steam is formed. The steam is condensed as water. What is the volume of gas remaining in the apparatus?
First, write the equation.
Does the reaction need equal volumes of hydrogen and oxygen?
Which gas will be left over after the reaction?
Calculate what volume of this gas is used during the reaction.
Subtract the volume used from the starting volume to find out what volume is left.

## Section 2

1. The complete combustion of carbon in oxygen yields carbon dioxide. Calculate the volume of oxygen at r.t.p. that would react with 10.0 g of carbon and the volume of carbon dioxide formed.

2. In the reaction between marble and hydrochloric acid, the equation is

$$CaCO_3(s) + 2HCl(aq) \rightarrow CaCl_2(aq) + CO_2(g) + H_2O(l)$$

What mass of marble would be needed to give 11.00 g of carbon dioxide?
What volume would this gas occupy at r.t.p.?

3. Zinc reacts with aqueous hydrochloric acid to give hydrogen.

$$Zn(s) + 2HCl(aq) \rightarrow H_2(g) + ZnCl_2(aq)$$

What mass of zinc would be needed to give 100 g of hydrogen? What volume would this gas occupy at r.t.p.?

4. Lead(II) oxide can be reduced by hydrogen to lead. What volume of hydrogen at r.t.p. would be needed to reduce 4.46 g of lead(II) oxide to lead? What mass of lead would be formed?

$$PbO(s) + H_2(g) \rightarrow Pb(s) + H_2O(g)$$

5. What volume of oxygen (at r.t.p.) would be formed by the complete thermal decomposition of 3.31 g of lead(II) nitrate? The equation for the reaction is

$$2Pb(NO_3)_2(s) \rightarrow 2PbO(s) + 4NO_2(g) + O_2(g)$$

6. What volume of oxygen (at r.t.p.) is formed by the decomposition of a solution containing 1.70 g of hydrogen peroxide?

$$2H_2O_2(aq) \rightarrow 2H_2O(l) + O_2(g)$$

7. What mass of potassium chlorate(V) must be heated to give 120 $cm^3$ of oxygen (at r.t.p.), according to the equation

$$2KClO_3(s) \rightarrow 2KCl(s) + 3O_2(g)$$

8. A mixture of hydrogen and oxygen can be exploded to form water. What volume of oxygen is needed to convert 123 $cm^3$ of hydrogen into water?

9. Propane burns in oxygen to form carbon dioxide and water:

$$C_3H_8(g) + 5O_2(g) \rightarrow 3CO_2(g) + 4H_2O(g)$$

What volume of oxygen (at r.t.p.) is required for the complete combustion of 44 g of propane? What volume of carbon dioxide is formed?

10. Sodium hydrogencarbonate decomposes on heating, with evolution of carbon dioxide:

$$2NaHCO_3(s) \rightarrow Na_2CO_3(s) + CO_2(g) + H_2O(g)$$

What volume of carbon dioxide (at r.t.p.) can be obtained by heating 4.20 g of sodium hydrogencarbonate? If 4.2 g of sodium hydrogencarbonate react with an excess of dilute hydrochloric acid, what volume of carbon dioxide (at r.t.p.) is evolved?

## Section 3

1. The problem is to find the percentage by volume composition of a mixture of hydrogen and ethane. When 75 $cm^3$ of the mixture was burned in an excess of oxygen, the volume of carbon dioxide produced was 60 $cm^3$ (all volumes at r.t.p.).

   (a) Write an equation for the combustion of ethane.

   (b) Say what volume of ethane would give 60 $cm^3$ of carbon dioxide.

   (c) Calculate the percentage of ethane in the mixture.

2. 25 $cm^3$ of carbon monoxide were ignited with 25 $cm^3$ of oxygen. All gas volumes were measured at the same temperature and pressure. The reduction in the total volume was

   **A** 2.5 $cm^3$
   **B** 10.0 $cm^3$
   **C** 12.5 $cm^3$
   **D** 15.0 $cm^3$
   **E** 25.0 $cm^3$

3. Ethene reacts with oxygen according to the equation

$$C_2H_4(g) + 3O_2(g) \rightarrow 2CO_2(g) + 2H_2O(l)$$

   15.0 $cm^3$ of ethene were mixed with 60.0 $cm^3$ of oxygen and the mixture was sparked to complete the reaction. If all the volumes were measured at r.t.p., the volume of the products would be:

   **A** 15 $cm^3$
   **B** 30 $cm^3$
   **C** 45 $cm^3$
   **D** 60 $cm^3$
   **E** 75 $cm^3$

4. When North Sea gas burns completely, it forms carbon dioxide and water and no other products. When 250 $cm^3$ of North Sea gas burn, they need 500 $cm^3$ of oxygen, and they form 250 $cm^3$ of carbon dioxide and 500 $cm^3$ of steam. Deduce the equation for the reaction and the formula of North Sea gas.

5. Ethene, $H_2C{=}CH_2$, and hydrogen react in the presence of a nickel catalyst to form ethane.

   (a) Write a balanced equation for the reaction.

(b) If a mixture of 30 $cm^3$ of ethene and 20 $cm^3$ of hydrogen is passed over a nickel catalyst, what is the composition of the final mixture? (Assume that the reaction is complete and that all gas volumes are at r.t.p.)

6. Octane burns completely in oxygen according to the equation

$$2C_8H_{18}(g) + 25O_2(g) \rightarrow 16CO_2(g) + 18H_2O(l)$$

(a) Calculate the volume of oxygen that is needed for the complete combustion of 1.00 $dm^3$ of octane vapour (all volumes at r.t.p.).

(b) What mass of water vapour will be formed?

7. Butane is used as a fuel in camping stoves. The equation for the complete combustion of butane is

$$2C_4H_{10}(g) + 13O_2(g) \rightarrow 8CO_2(g) + 10H_2O(l)$$

100 $cm^3$ of butane are burned in an excess of oxygen. Calculate:

(a) the volume of oxygen used

(b) the volume of carbon dioxide produced. (Assume all gas volumes are at r.t.p.)

8. The equation for the combustion of ethane is

$$2C_2H_6(g) + 7O_2(g) \rightarrow 4CO_2(g) + 6H_2O(l)$$

90 g of ethane were burned completely in oxygen.

(a) What volume of carbon dioxide (at r.t.p.) was formed?

(b) How many moles of oxygen were used up?

(c) What volume of air (20% by volume of oxygen) would be needed for the complete combustion of 90 g of ethane (at r.t.p.)?

9. The table gives the formulae and relative molecular masses of some gases.

| Formula | Ne | $C_2H_2$ | $O_2$ | Ar | $NO_2$ | $SO_2$ | $SO_3$ |
|---|---|---|---|---|---|---|---|
| $M_r$ | 20 | 26 | 32 | 40 | 46 | 64 | 80 |
| Volume ($cm^3$) occupied by 1 g of gas at r.t.p. | 1200 | 923 | 750 | 600 | 520 | 375 | 300 |

(a) Plot a graph of volume (on the vertical axis) against $M_r$ (on the horizontal axis).

(b) Use the graph to predict the volumes occupied at r.t.p. by
(i) 1 g of fluorine, $F_2$
(ii) 1 g of $Cl_2$.

(c) What is the relative molecular mass of a gas which occupies 545 $cm^3$ per gram at r.t.p.? If the gas contains only carbon and oxygen, what is its formula?

## Section 4 Problems involving both Masses of Solids and Volumes of Gases

1. A certain industrial plant emits 90 tonnes of nitrogen monoxide, NO, daily through its chimneys. The firm decides to remove nitrogen monoxide from its waste gases by means of the reaction

   methane + nitrogen monoxide → nitrogen + carbon dioxide + water

   $$CH_4(g) + 4NO(g) \rightarrow 2N_2(g) + CO_2(g) + 2H_2O(l)$$

   If methane (North Sea gas) costs 0.50 p per cubic metre, what will this clean-up reaction cost the firm to run? (Ignore the cost of installing the process, which will in reality be high.)

   [*Hint*: tonnes of NO . . . moles of NO . . . moles of $CH_4$ . . . volume of $CH_4$ . . . cost of $CH_4$ ($1\ m^3 = 1000\ dm^3$)]

2. When magnesium nitride reacts with water, the reaction which takes place is:

   magnesium nitride + water → magnesium hydroxide + ammonia

   $$Mg_3N_2(s) + 6H_2O(l) \rightarrow 3Mg(OH)_2(s) + 2NH_3(g)$$

   (a) How many moles of $Mg(OH)_2$ are produced from 0.100 mole of $Mg_3N_2$?

   (b) What volume of $NH_3$ (at r.t.p.) is produced from the reaction of 0.100 mole of $Mg_3N_2$?

   (c) If 500 g of $Mg_3N_2$ react with an excess of water, how many moles of
   (i) $Mg(OH)_2$ and (ii) $NH_3$ are formed?

   (d) What mass of $Mg_3N_2$ is needed to produce 0.075 mole of $Mg(OH)_2$?

   (e) What is the mass of $Mg_3N_2$ that is needed to produce 116 g of $Mg(OH)_2$?

3. (a) Analysis of an oxide of potassium shows that 1.42 g of this oxide contains 0.64 g of oxygen. What is its empirical formula?

   (b) This oxide reacts with carbon dioxide to form oxygen and potassium carbonate, $K_2CO_3$. Write an equation for the reaction.

   (c) This reaction is sometimes used as a means of regenerating oxygen in submarines. What volume of oxygen (at r.t.p.) could be obtained from 1.00 kg of this oxide of potassium?

4. A cook is making a small cake. It needs 500 $cm^3$ (at r.t.p.) of carbon dioxide to make the cake rise. The cook decides to add baking powder, which contains sodium hydrogencarbonate. This generates carbon dioxide by thermal decomposition:

   $$2NaHCO_3(s) \rightarrow CO_2(g) + Na_2CO_3(s) + H_2O(l)$$

What mass of sodium hydrogencarbonate must the cook add to the cake mixture?

5. Dinitrogen oxide, $N_2O$, is commonly called *laughing gas*. It can be made by heating ammonium nitrate, $NH_4NO_3$. The *unbalanced* equation for this reaction is:

$$NH_4NO_3(s) \rightarrow N_2O(g) + H_2O(l)$$

(a) Balance the equation.

(b) Calculate the mass of ammonium nitrate that must be heated to give
   (i) 8.8 g of laughing gas
   (ii) 10.0 $dm^3$ of laughing gas (at r.t.p.).

6. Many years ago, bicycle lamps used to burn the gas ethyne, $C_2H_2$. The gas was produced by allowing water to drip on to calcium carbide. The *unbalanced* equation for the reaction is:

calcium carbide + water → ethyne + calcium hydroxide

$$CaC_2(s) + H_2O(l) \rightarrow C_2H_2(g) + Ca(OH)_2(s)$$

(a) Balance the equation.

(b) Calculate the mass of calcium carbide which would be needed to produce 500 $cm^3$ of ethyne (at r.t.p.).

7. An excess of magnesium carbonate, $MgCO_3$, was added to 50.0 $cm^3$ of 1.00 mol/$dm^3$ sulphuric acid. Calculate:

(a) the mass of magnesium carbonate that reacted

(b) the volume at r.t.p. of carbon dioxide evolved.

8. Methane can be made by the action of water on aluminium carbide:

aluminium carbide + water → methane + aluminium hydroxide

$$Al_4C_3(s) + 12H_2O(l) \rightarrow 3CH_4(g) + 4Al(OH)_3(s)$$

What volume of methane (at r.t.p.) can be obtained from 0.36 g of aluminium carbide?
[*Hint*: mass of aluminium carbide . . . moles of aluminium carbide . . . moles of methane . . . volume of methane]

9. What volume of hydrogen (at r.t.p.) can be obtained when 2.6 g of zinc reacts with an excess of dilute sulphuric acid?

10. What volume of ammonia (at r.t.p.) can be obtained by heating 0.25 mol of ammonium sulphate with calcium hydroxide?

$$(NH_4)_2SO_4(s) + Ca(OH)_2(s) \rightarrow 2NH_3(g) + CaSO_4(s) + 2H_2O(l)$$

# 6. The Volumes of Solutions which React Together

## Concentration

One way of stating the concentration of a solution is to state the *mass* of solute present in 1 cubic decimetre of solution. The mass of solute is usually expressed in grams. A solution made by dissolving 5 grams of solute and making up to 1 cubic decimetre of solution has a concentration of $5 g/dm^3$ (5 grams per cubic decimetre).

Other units of volume are the cubic centimetre, $cm^3$, the cubic metre, $m^3$, and the litre, l. The litre has the same volume as the cubic decimetre.

$$10^3 \text{ cm}^3 = 1 \text{ dm}^3 = 1 \text{ l} = 10^{-3} \text{ m}^3 \quad (10^3 = 1000;\ 10^{-3} = 0.001)$$

A more common way of stating concentration in chemistry is to state the **molar concentration** of a solution. This is the **amount in moles** of a substance present in 1 cubic decimetre (1 litre) of solution.

What is the *molar concentration* of a solution of 80 g of sodium hydroxide in 1 $dm^3$ of solution? The number of moles of NaOH in 80 g of sodium hydroxide can be calculated from its relative molecular mass.

Relative molecular mass of NaOH = 23 + 16 + 1 = 40

$$\text{Number of moles of substance} = \frac{\text{Mass of substance}}{\text{Mass of 1 mole of substance}}$$

$$\text{Number of moles of NaOH} = \frac{80 \text{ g}}{40 \text{ g}} = 2$$

*Another way of saying the number of moles of sodium hydroxide is 2 is to say the amount of sodium hydroxide is 2 moles.*

The concentration of the solution is given by:

$$\text{Concentration of solution (mol/dm}^3\text{)} = \frac{\text{Amount of solute (mol)}}{\text{Volume of solution (dm}^3\text{)}}$$

For this solution,

$$\text{concentration} = \frac{2 \text{ mol}}{1 \text{ dm}^3} = 2 \text{ mol/dm}^3$$

If 3 moles of sodium hydroxide are dissolved in 500 $cm^3$ of solution,

$$\text{concentration} = \frac{3 \text{ mol}}{0.5 \text{ dm}^3} = 6 \text{ mol/dm}^3$$

The symbol M is often used for mol/$dm^3$. This solution can be described as a 6 M sodium hydroxide solution.

Figure 6.1 gives more examples.

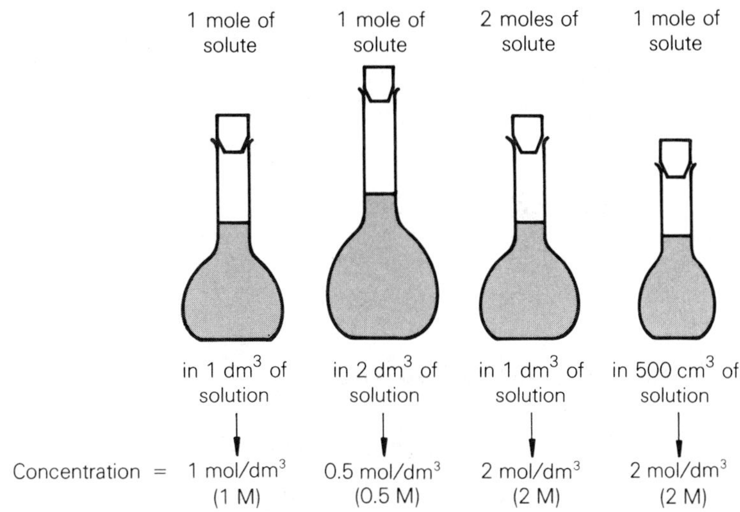


Figure 6.1 How to calculate concentration

A useful rearrangement of the expression in the box is:

Amount of solute (mol) = Volume ($dm^3$) × Concentration (mol/$dm^3$)

For example, the amount in moles of solute in 2.5 $dm^3$ of a 2.0 mol/$dm^3$ solution is given by:

Amount of solute = 2.5 $dm^3$ × 2.0 mol/$dm^3$ = 5.0 mol

**Example 1** Calculate the concentration in mol/$dm^3$ of a solution containing 36.5 g of hydrogen chloride in 4.00 $dm^3$ of solution.

**Method:**

Relative molecular mass of HCl = (35.5 + 1.0) = 36.5

Amount in moles present in 36.5 g = 1.00 mol

Volume = 4.00 $dm^3$

Concentration of solution $= \dfrac{\text{Amount of solute in moles}}{\text{Volume of solution in dm}^3}$

$= \dfrac{1.00 \text{ mol}}{4.00 \text{ dm}^3}$

$= 0.25 \text{ mol/dm}^3$

**Answer:** The concentration is 0.25 mol/$dm^3$. (It is a 0.25 M solution.)

**Example 2** Calculate the amount of solute in moles in 250 $cm^3$ of a solution of sodium hydroxide which has a concentration of 2.00 $mol/dm^3$.

**Method:**

Concentration of solution = 2.00 $mol/dm^3$

Volume of solution = 250 $cm^3$ = 0.250 $dm^3$

Amount of solute (mol) = Volume ($dm^3$) × Concentration ($mol/dm^3$)

= 2.00 × 0.250 = 0.500 mol

**Answer:** The solution contains 0.500 moles of solute.

# Problems on Concentration

1. Calculate the concentration in $mol/dm^3$ of

(a) 3.65 g of hydrogen chloride in 2.00 $dm^3$ of solution

(b) 73.0 g of hydrogen chloride in 2.00 $dm^3$ of solution

(c) 49.0 g of sulphuric acid in 2.00 $dm^3$ of solution

(d) 49.0 g of sulphuric acid in 250 $cm^3$ of solution

(e) 2.80 g of potassium hydroxide in 500 $cm^3$ of solution

(f) 28.0 g of potassium hydroxide in 4.00 $dm^3$ of solution

(g) 5.30 g of anhydrous sodium carbonate in 200 $cm^3$ of solution

(h) 53.0 g of anhydrous sodium carbonate in 2.50 $dm^3$ of solution

2. Calculate the amount in moles of solute in

(a) 250 $cm^3$ of sodium hydroxide solution containing 1.00 $mol/dm^3$

(b) 500 $cm^3$ of sodium hydroxide solution containing 0.250 $mol/dm^3$

(c) 250 $cm^3$ of 0.0200 M calcium hydroxide solution

(d) 2.00 $dm^3$ of 1.25 M sulphuric acid (1.25 $mol/dm^3$)

(e) 125 $cm^3$ of aqueous nitric acid, having a concentration of 0.400 $mol/dm^3$

(f) 200 $cm^3$ of ammonia solution, having a concentration of 0.125 $mol/dm^3$

(g) 123 $cm^3$ of aqueous hydrochloric acid of concentration 3.00 $mol/dm^3$

(h) 1500 $cm^3$ of potassium hydroxide solution of concentration 0.750 $mol/dm^3$.

# Neutralisation

When hydrochloric acid is neutralised by sodium hydroxide, the equation is

$$HCl(aq) + NaOH(aq) \rightarrow NaCl(aq) + H_2O(l)$$

1 mole of hydrochloric acid needs 1 mole of sodium hydroxide.
1 mole of HCl is present in 1 $dm^3$ of solution of concentration 1 mol/$dm^3$.
1 mole of NaOH is present in 1 $dm^3$ of solution of concentration 1 mol/$dm^3$.
1 $dm^3$ of hydrochloric acid of concentration 1 mol/$dm^3$ therefore neutralises 1 $dm^3$ of sodium hydroxide of concentration 1 mol/$dm^3$.

Does 1 $dm^3$ of acid of concentration 1 mol/$dm^3$ always neutralise 1 $dm^3$ of alkali of concentration 1 mol/$dm^3$? When sulphuric acid is neutralised by sodium hydroxide solution

$$H_2SO_4(aq) + 2NaOH(aq) \rightarrow Na_2SO_4(aq) + 2H_2O(l)$$

1 mole of sulphuric acid neutralises 2 moles of sodium hydroxide, and, therefore, 1 $dm^3$ of sulphuric acid of concentration 1 mol/$dm^3$ neutralises 2 $dm^3$ of sodium hydroxide solution of concentration 1 mol/$dm^3$.

In the reaction between hydrochloric acid and sodium carbonate solution

$$2HCl(aq) + Na_2CO_3(aq) \rightarrow 2NaCl(aq) + CO_2(g) + H_2O(l)$$

2 moles of hydrochloric acid are needed by 1 mole of sodium carbonate, and 2 $dm^3$ of hydrochloric acid of concentration 1 mol/$dm^3$ neutralise 1 $dm^3$ of sodium carbonate solution of concentration 1 mol/$dm^3$.

When you are doing calculations on reacting volumes of solutions, the equation for the reaction is important. Look at the equation to find out how many moles of acid react with one mole of alkali. The problems in Section 1 on page 48 will give you practice at this.

# Titration

It is possible to find out the concentration of a solution by finding out what volume of it will react with a known volume of a solution of known concentration. For example, you can find out the concentration of an alkaline solution by finding out what volume of it will neutralise, say, 25 $cm^3$ of a solution of an acid of known concentration. The procedure of adding one liquid to another in a measured way is called **titration**. A solution of known concentration is called a **standard solution**.

**Example 1** In a titration of a solution of unknown concentration sodium hydroxide against standard hydrochloric acid, 25.0 $cm^3$ of sodium hydroxide solution neutralise 22.5 $cm^3$ of aqueous hydrochloric acid of concentration 0.90 mol/$dm^3$. What is the concentration in mol/$dm^3$ of the sodium hydroxide solution?

**Method:** In tackling this calculation,

(a) Find out the number of moles of acid needed to neutralise one mole of alkali.

(b) Use the expression

Amount of solute (mol) = Volume ($dm^3$) × Concentration ($mol/dm^3$)

Amount of hydrochloric acid (mol) = Volume ($dm^3$) × Concentration ($mol/dm^3$)

$= 22.5 \times 10^{-3} \times 0.90$ mol

Since

$$NaOH(aq) + HCl(aq) \rightarrow NaCl(aq) + H_2O(l)$$

Number of moles of hydrochloric acid = Number of moles of sodium hydroxide

Therefore, amount of sodium hydroxide (mol) $= 22.5 \times 10^{-3} \times 0.90$

But, amount of sodium hydroxide (mol) = Volume ($dm^3$) × Concentration ($mol/dm^3$)

$= 25.0 \times 10^{-3} \times c$

where $c$ = concentration

$$25.0 \times 10^{-3} \times c = 22.5 \times 10^{-3} \times 0.90$$

$$c = \frac{22.5 \times 10^{-3} \times 0.90}{25.0 \times 10^{-3}}$$

$$= 0.81 \text{ mol/dm}^3$$

**Answer:** The sodium hyroxide solution contains 0.81 $mol/dm^3$.

**Example 2** Titration shows that 25.0 $cm^3$ of sodium hydroxide solution neutralise 22.5 $cm^3$ of sulphuric acid of concentration 0.90 $mol/dm^3$. What is the concentration of the sodium hydroxide solution?

**Method:** As in Example 1:

(a) Find the number of moles of acid needed to neutralise 1 mole of alkali.

(b) Amount of solute (mol) = Volume ($dm^3$) × Concentration ($mol/dm^3$)

Amount of sulphuric acid (mol) = Volume ($dm^3$) × Concentration ($mol/dm^3$)

$= 22.5 \times 10^{-3} \times 0.90$ mol

Since

$$2NaOH(aq) + H_2SO_4(aq) \rightarrow Na_2SO_4(aq) + 2H_2O(l)$$

Number of moles of sodium hydroxide $= 2 \times$ Number of moles of sulphuric acid

$= 2 \times 22.5 \times 10^{-3} \times 0.90$

But, amount of sodium hydroxide (mol) $=$ Volume $\times$ Concentration

$= 25.0 \times 10^{-3} \times c$

where $c =$ concentration

Therefore, $25.0 \times 10^{-3} \times c = 2 \times 22.5 \times 10^{-3} \times 0.90$

$$c = \frac{2 \times 22.5 \times 10^{-3} \times 0.90}{25.0 \times 10^{-3}}$$

$= 1.62$ mol/dm$^3$

**Answer:** The concentration of the sodium hydroxide solution is 1.62 mol/dm$^3$.

*Note* that, although the titration figures in this example are the same as those in Example 1, the concentration of sodium hydroxide is twice what it was calculated to be in Example 1 because, in Example 2, 1 mole of acid neutralises 2 moles of alkali.

**Example 3** What volume of hydrochloric acid of concentration 0.25 mol/dm$^3$ (0.25 M) is needed to neutralise 5.3 g of anhydrous sodium carbonate?

**Method:** You tackle this example, as before, by finding the number of moles of each substance. For the solid, use the expression:

$$\text{Number of moles of solid} = \frac{\text{Mass of solid}}{\text{Mass of 1 mole of solid}}$$

For the solution, use the expression:

Amount of solute (mol) $=$ Volume (dm$^3$) $\times$ Concentration (mol/dm$^3$)

Relative molecular mass of $Na_2CO_3$ $= (2 \times 23) + 12 + (3 \times 16) = 106$

Number of moles of $Na_2CO_3$ $=$ Mass/$M_r$ $= 5.3/106$

$= 1/20 = 0.05$

Since

$$Na_2CO_3(s) + 2HCl(aq) \rightarrow 2NaCl(aq) + CO_2(g) + H_2O(l)$$

Number of moles of HCl $= 2 \times$ number of moles of $Na_2CO_3$

$= 2 \times 0.05 = 0.10$

But amount of HCl in solution (mol) $=$ Volume (dm$^3$) $\times$ Concentration (mol/dm$^3$)

Therefore $0.10 =$ Volume $\times 0.25$

Volume $= 0.10/0.25 = 0.4$ dm$^3$

**Answer:** The volume needed is 0.4 dm$^3$ or 400 cm$^3$.

# Problems on Reacting Volumes of Solutions

## Section 1

Calculators are not needed for these problems.

The following are problems on neutralisation. Show, giving your working, whether each of these statements is true or false.

1. 1 mole of HCl will neutralise
   (a) 5 $dm^3$ of KOH(aq) of concentration 0.2 mol/$dm^3$: True or False?
   (b) 2 $dm^3$ of NaOH(aq) of concentration 0.2 mol/$dm^3$
   (c) 2 $dm^3$ of KOH(aq) of concentration 0.5 mol/$dm^3$
   (d) 0.5 $dm^3$ of NaOH(aq) of concentration 1 mol/$dm^3$
   (e) 250 $cm^3$ of $Na_2CO_3$(aq) of concentration 2 mol/$dm^3$
   (f) 200 $cm^3$ of $Na_2CO_3$(aq) of concentration 4 mol/$dm^3$

2. 1 mole of $H_2SO_4$ will neutralise
   (a) 500 $cm^3$ of NaOH(aq) of concentration 4 mol/$dm^3$: True or False?
   (b) 1 $dm^3$ of KOH(aq) of concentration 1 mol/$dm^3$
   (c) 400 $cm^3$ of NaOH(aq) of concentration 5 mol/$dm^3$
   (d) 500 $cm^3$ of $Na_2CO_3$(aq) of concentration 1 mol/$dm^3$
   (e) 2 $dm^3$ of $Na_2CO_3$(aq) of concentration 0.5 mol/$dm^3$
   (f) 4 $dm^3$ of KOH(aq) of concentration 0.25 mol/$dm^3$

3. 5 moles of NaOH will neutralise
   (a) 2 $dm^3$ of HCl(aq) of concentration 2 mol/$dm^3$: True or False?
   (b) 250 $cm^3$ of HCl(aq) of concentration 10 mol/$dm^3$
   (c) 250 $cm^3$ of $H_2SO_4$(aq) of concentration 10 mol/$dm^3$
   (d) 500 $cm^3$ of $H_2SO_4$(aq) of concentration 5 mol/$dm^3$
   (e) 2500 $cm^3$ of $HNO_3$(aq) of concentration 2 mol/$dm^3$
   (f) 2 $dm^3$ of $HNO_3$(aq) of concentration 2 mol/$dm^3$

4. 0.5 mole of $Na_2CO_3$ will neutralise
   (a) 1 $dm^3$ of HCl(aq) of concentration 1 mol/$dm^3$: True or False?
   (b) 1 $dm^3$ of $H_2SO_4$(aq) of concentration 1 mol/$dm^3$
   (c) 500 $cm^3$ of HCl(aq) of concentration 1 mol/$dm^3$
   (d) 250 $cm^3$ of $HNO_3$(aq) of concentration 2 mol/$dm^3$
   (e) 200 $cm^3$ of $H_2SO_4$(aq) of concentration 2.5 mol/$dm^3$
   (f) 500 $cm^3$ of $HNO_3$(aq) of concentration 2 mol/$dm^3$

## Section 2

These problems can be solved without the use of calculators.

1. 25.0 $cm^3$ of a solution of potassium hydroxide are neutralised by 35.0 $cm^3$ of aqueous hydrochloric acid of concentration 0.75 mol/$dm^3$. Calculate the concentration in mol/$dm^3$ of the potassium hydroxide solution.

2. If 25.0 $cm^3$ of a solution of sodium hydroxide are neutralised by 27.5 $cm^3$ of aqueous sulphuric acid of concentration 0.250 mol/$dm^3$, what is the concentration (mol/$dm^3$) of the sodium hydroxide solution?

3. Sodium reacts with water according to the equation

$$2Na(s) + 2H_2O(l) \rightarrow 2NaOH(aq) + H_2(g)$$

Find the volume of aqueous hydrochloric acid, of concentration 0.25 mol/$dm^3$, needed to neutralise the sodium hydroxide formed by the reaction of 0.23 g of sodium.

4. A solution of hydrochloric acid was titrated against a solution of sodium carbonate of concentration 0.250 mol/$dm^3$. 25.0 $cm^3$ of the acid solution neutralised 20.0 $cm^3$ of the alkali. Calculate
   (a) the number of moles of alkali used in the titration
   (b) the number of moles of acid used in the titration
   (c) the concentration of the acid in mol/$dm^3$.

5. 2.1 g of sodium hydrogencarbonate react with dilute hydrochloric acid with the evolution of carbon dioxide. What volume of a solution of hydrochloric acid containing 0.50 mol/$dm^3$ is needed to liberate the maximum possible volume of carbon dioxide? What volume of carbon dioxide (at r.t.p.) is obtained?

6. A solution of sodium carbonate contains 53.0 g in 250 $cm^3$ of solution. Calculate
   (a) the concentration (mol/$dm^3$) of the sodium carbonate solution
   (b) the volume of aqueous hydrochloric acid of concentration 0.250 mol/$dm^3$ needed to neutralise 25.0 $cm^3$ of the sodium carbonate solution.

7. Hydrogen is formed by the reaction of a solution of sulphuric acid with an excess of magnesium. What volume of aqueous sulphuric acid, of concentration 1.5 mol/$dm^3$, must be used to give 1200 $cm^3$ of hydrogen (at r.t.p.)?

## Section 3

1. Recently a chain of American cafeterias started giving away brightly painted drinking glasses. It was found that lead compounds in the

paint dissolved in acidic drinks, such as fruit juices. In one test, 100 $cm^3$ of juice were found to contain 4.0 mg of lead. What was the concentration of lead in

(a) $g/dm^3$ (b) $mol/dm^3$?

2. Sodium hydroxide is sold commercially as solid **lye**. A 1.20 g sample of lye required 45.0 $cm^3$ of 0.500 $mol/dm^3$ hydrochloric acid to neutralise it. Calculate the percentage by mass of NaOH in lye.

3. Vinegar is a solution of ethanoic acid. A 10.0 $cm^3$ portion of a certain brand of vinegar needed 55.0 $cm^3$ of 0.200 $mol/dm^3$ sodium hydroxide solution to neutralise the ethanoic acid in it.

ethanoic acid + sodium hydroxide → sodium ethanoate + water

$$CH_3CO_2H(aq) + NaOH(aq) \rightarrow CH_3CO_2Na(aq) + H_2O(l)$$

(a) Calculate the concentration of ethanoic acid in the vinegar in $mol/dm^3$.

(b) Given that the density of this vinegar is 1.06 $g/cm^3$, calculate the concentration of ethanoic acid in percentage by mass.

4. Salt is a necessary ingredient of our diet. In certain illnesses, the salt balance can be lost, and a doctor or nurse must give salt intravenously. They inject **normal saline**, which is a 0.85% solution of sodium chloride in water (0.85 g of solute per 100 g of water). What is the molar concentration of normal saline?

5. A chip of marble weighing 2.50 g required 28.0 g of 1.50 $mol/dm^3$ hydrochloric acid to react with all the calcium carbonate it contained. What is the percentage of calcium carbonate in this sample of marble?

(a) Write the balanced equation for the reaction.

(b) Find how many moles of HCl were used . . . then how many moles of $CaCO_3$ reacted . . . what mass of $CaCO_3$ . . . and finally the percentage of $CaCO_3$.

6. In an adult, the concentration of potassium ions in the blood ranges from $3.5 \times 10^{-3}$ $mol/dm^3$ to $5.3 \times 10^{-3}$ $mol/dm^3$. When a patient taking a diuretic has a blood test, the analyst finds that a 50 $cm^3$ blood sample contains $3.9 \times 10^{-3}$ g of potassium ions. Does the patient need to be given a potassium supplement?

7. The blood of a patient who is being treated with lithium carbonate for depression is sampled. It is found to contain 1.4 mg of lithium in 100 $cm^3$. The concentration of lithium ions in blood should not exceed $1.5 \times 10^{-3}$ $mol/dm^3$. Should the patient stop taking lithium carbonate? Explain your answer.

8. Magnesium reacted with dilute sulphuric acid according to the equation

$$Mg(s) + H_2SO_4(aq) \rightarrow H_2(g) + MgSO_4(aq)$$

50 $cm^3$ of 2.0 $mol/dm^3$ sulphuric acid were treated with an excess of magnesium.

(a) What mass of magnesium reacted?

(b) What volume of hydrogen (at r.t.p.) was formed?

9. A mixture of gases coming from a coke-producing plant contains ammonia. The mixture is bubbled through dilute sulphuric acid to remove the ammonia.

(a) Write a balanced equation for the reaction which occurs.

(b) What volume of ammonia (at r.t.p.) could be removed by 50 $dm^3$ of 1.50 $mol/dm^3$ sulphuric acid?

(c) What use could be made of the product?

10. Nitrosoamines can cause cancer at sufficiently high concentrations. In 1979, a brand of American beer was found to contain 6 p.p.b (parts per billion) of dimethylnitrosoamine. By 1981, the firm had reduced the level to 0.2 p.p.b.

(a) What was the mass of dimethylnitrosoamine in one 250 $cm^3$ can of beer in 1979? (1 billion = $10^9$)

(b) What fraction of the 1979 level was still present in 1981?

11. If a person's blood sugar level falls below 60 mg per 100 $cm^3$, insulin shock can occur. The density of blood is 1.2 $g/cm^3$.

(a) What is the percentage by mass of sugar in the blood at this level?

(b) What is the molar concentration of sugar, $C_6H_{12}O_6$, in the blood?

12. A blood alcohol level of 150-200 mg alcohol per 100 $cm^3$ of blood produces intoxication. A blood alcohol level 300-400 mg per 100 $cm^3$ produces unconsciousness. At a blood alcohol level above 500 mg per 100 $cm^3$, a person may die. What is the molar concentration of alcohol (ethanol, $C_2H_5OH$) at the lethal level?

13. An experiment was done to find the percentage composition of an alloy of sodium and lead. The alloy reacts with water:

$$\text{alloy} + \text{water} \rightarrow \text{sodium hydroxide} + \text{hydrogen} + \text{lead}$$

$$2Na.Pb(s) + 2H_2O(l) \rightarrow 2NaOH(aq) + H_2(g) + 2Pb(s)$$

3.00 g of the alloy were added to about 100 $cm^3$ of water. When the reaction was complete, the sodium hydroxide formed was titrated against 1.00 $mol/dm^3$ hydrochloric acid. The volume of acid required to neutralise the sodium hydroxide was 12.0 $cm^3$. Calculate

(a) the amount in moles of HCl used

(b) the amount in moles of NaOH neutralised

(c) the amount in moles of Na in 3.00 g of the alloy

(d) the mass in grams of Na in 3.00 g of alloy

(e) the percentage composition by mass.

14. 100 $cm^3$ of 1.0 mol/$dm^3$ ammonia solution was neutralised by a certain volume of 1.0 mol/$dm^3$ sulphuric acid. The resulting solution was evaporated and allowed to crystallise. Crystals of ammonium sulphate formed.

$$2NH_3(aq) + H_2SO_4(aq) \rightarrow (NH_4)_2SO_4(aq)$$

(a) What volume of 1.0 mol/$dm^3$ sulphuric acid was required to neutralise 100 $cm^3$ of 1.0 mol/$dm^3$ ammonia?

(b) What mass of ammonium sulphate was formed in (a)?

(c) Explain why the mass of crystals obtained from the solution is less than the calculated quantity.

15. Ammonia is manufactured by the Haber process. A 500 $cm^3$ sample of the gases leaving the catalyst chamber was bubbled into 0.100 mol/$dm^3$ sulphuric acid. 20.0 $cm^3$ of the acid were neutralised. What was the percentage by volume of ammonia in the sample? Name two other gases which were present.

16. Solid calcium hydroxide is shaken with water at room temperature until a saturated solution has been formed. A titration is carried out to find the concentration of the solution. 25.0 $cm^3$ of the saturated solution are neutralised by 10.5 $cm^3$ of 0.100 mol/$dm^3$ hydrochloric acid.

Calculate the solubility of calcium hydroxide in
(a) mol/$dm^3$ (b) g/$dm^3$.

17. Temporary hardness in tapwater is caused by the presence of calcium hydrogencarbonate. This solute can be estimated by titration against standard hydrochloric acid.

$$Ca(HCO_3)_2(aq) + 2HCl(aq) \rightarrow CaCl_2(aq) + 2CO_2(g) + 2H_2O(l)$$

One litre of tapwater required 12.0 $cm^3$ of 0.500 mol/l hydrochloric acid in a titration. Find the concentration of $Ca(HCO_3)_2$ in tapwater in

(a) mol/l (b) g/l.

# 7. Heat Changes in Chemical Reactions

## Heat Changes

There are two types of chemical reactions. In one, the products contain less energy than the reactants; in the other, the products contain more energy than the reactants. The two types of reactions can be represented by the energy diagrams shown in Figure 7.1. The difference between the energy of the reactants and the energy of the products is called the **heat of reaction** and is represented as $\boldsymbol{\Delta H}$.

*Heat of reaction* $\Delta H$ = *Energy of products* − *Energy of reactants*

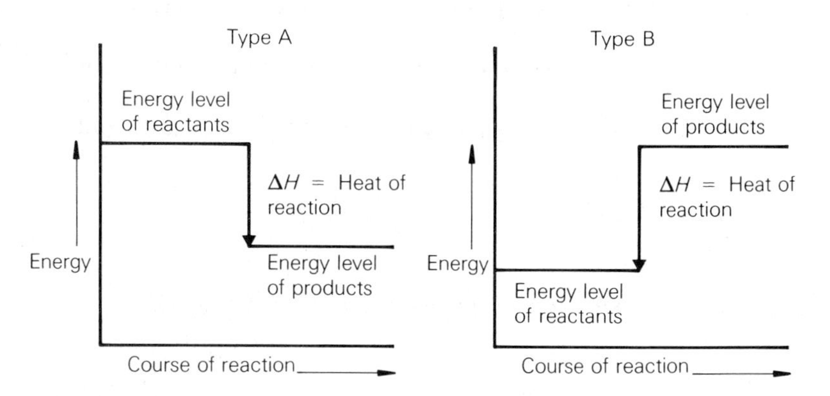


Figure 7.1 Energy changes during chemical reactions

In reactions of Type A, since the products possess less energy than the reactants, $\Delta H$ is negative. The way the reactants get rid of the excess energy is by giving out heat, and this type of reaction is **exothermic**. In reactions of Type B, since the products have more energy than the reactants, $\Delta H$ is positive. Since the reactants must gain energy in order to form the products, the reaction takes in heat: it is **endothermic**.

Many types of reaction involve a heat change. For each type of reaction the **molar heat of reaction** is defined as the heat change per mole of reactant. Heat is measured in joules (J) and kilojoules (kJ). 1000 J = 1 kJ. **Molar heat of neutralisation** is the heat change that occurs when 1 mole of hydrogen ions is neutralised by a base:

$$H^+(aq) + OH^-(aq) \rightarrow H_2O(l);\ \Delta H = -57.2 \text{ kJ/mol}$$

**Molar heat of combustion** is the heat change that occurs when 1 mole of a substance is completely burnt in oxygen:

$2C_2H_6(g) + 7O_2(g) \rightarrow 4CO_2(g) + 6H_2O(l)$; $\Delta H = -1560$ kJ per mole of ethane

**Molar heat of reaction** is the heat change that occurs in a reaction between the number of moles of reactants indicated by the equation

$4Al(s) + 3O_2(g) \rightarrow 2Al_2O_3(s)$; $\Delta H = -3360$ kJ per mole of the reaction as written.

**Example 1** If the molar heat of combustion of propane = −2220 kJ/mol, calculate the heat given out when 1000 g of propane are completely burnt.

**Method:**
Relative molecular mass of $C_3H_8 = (3 \times 12) + (8 \times 1) = 44$

1 mole of propane has a mass of 44 g

When 44 g propane are burnt, the heat given out = 2220 kJ

When 1 g of propane is burnt, the heat given out $= \frac{2220}{44}$ kJ

When 1000 g propane are burnt, the heat given out

$= \frac{1000 \times 2220}{44}$ kJ

= 50 454 kJ.

**Answer:** The heat given out is 50 500 kJ.

**Example 2** When 100 $cm^3$ of 1.00 mol/$dm^3$ sodium hydroxide solution are added to 100 $cm^3$ of 1.00 mol/$dm^3$ hydrochloric acid, 5.7 kJ of heat are given out. Calculate the molar heat of neutralisation.

**Method:** Since 100 $cm^3$ of 1.00 mol/$dm^3$ NaOH(aq) contain 0.100 mol of $OH^-$(aq) ions,
and 100 $cm^3$ of 1.00 mol/$dm^3$ HCl(aq) contain 0.100 mol of $H^+$(aq) ions,
the two solutions neutralise each other to produce 0.100 mol of $H_2O$ molecules.

$$H^+(aq) + OH^-(aq) \rightarrow H_2O(l)$$

Heat given out when 0.100 mol of $H_2O$ forms = 5.7 kJ
Heat given out when 1.00 mol of $H_2O$ forms = 57 kJ

**Answer:** The molar heat of neutralisation = −57 kJ/mol.

# Problems on Heat of Reaction

## Section 1

No calculators are needed for this section.

1. If 2.9 g of ethanol burn with the evolution of 87 kJ of heat, what is the molar heat of combustion of ethanol?

2. The molar heat of combustion of propan-1-ol is −2010 kJ/mol. What mass of propanol must be burned to give out 402 kJ of energy?

3. The molar heat of combustion of sucrose ($C_{12}H_{22}O_{11}$) is −5640 kJ/mol. What mass of sucrose must be burnt to give out 1410 kJ of energy?

4. The molar heat of combustion of glucose ($C_6H_{12}O_6$) is −2820 kJ/mol. What mass of glucose must be burnt to produce 705 kJ of energy?

5. If 13.5 g of phenylmethanol ($C_7H_7OH$) burn to give 505 kJ of heat, what is the molar heat of combustion of phenylmethanol?

6. The equation for the reaction occurring when pentane is burnt is

$$C_5H_{12}(g) + 8O_2(g) \rightarrow 5CO_2(g) + 6H_2O(l); \Delta H = -3500 \text{ kJ}$$

Which of the following amounts of pentane, burnt completely in oxygen, will release 3500 kJ?

**A** 1 molecule

**B** 1 $dm^3$

**C** 1 kilogram

**D** 1 mole

**E** 1 gram

## Section 2

1. Methane, $CH_4$, and butane, $C_4H_{10}$, are both used as fuels. Their heats of combustion are $CH_4$ = −890 kJ/mol; $C_4H_{10}$ = −2880 kJ/mol.

(a) What does the − sign mean?

(b) Calculate the energy liberated by the combustion of
  (i) 1 kg of methane
  (ii) 1 kg of butane.

2. Petrol is a mixture of hydrocarbons. One of the major constituents is octane, $C_8H_{18}$. When octane burns completely in oxygen, the equation is:

$$2C_8H_{18}(g) + 25O_2(g) \rightarrow 16CO_2(g) + 18H_2O(l)$$

(a) Calculate the volume of oxygen which is needed to burn completely 1.00 litre of octane vapour.

(b) When 1.00 mole of octane vapour is burnt completely, 5520 kJ of heat are evolved. Calculate the heat evolved when 1.00 litre of octane vapour is burnt completely.

(c) What mass of water is produced?

3. When water is formed from its elements,

$$2H_2(g) + O_2(g) \rightarrow 2H_2O(l); \Delta H = -570 \text{ kJ/mol}$$

570 kJ is the quantity of heat which is

**A** gained from the surroundings when 1 mole of water is formed

**B** lost to the surroundings when 1 mole of water is formed

**C** lost to the surroundings when 1 mole of hydrogen burns

**D** lost to the surroundings when 2 moles of water are formed

**E** gained from the surroundings when 1 mole of hydrogen burns

4. Both methane, $CH_4$, and ethyne, $C_2H_2$, are used as fuels.

$$CH_4(g) + 2O_2(g) \rightarrow CO_2(g) + 2H_2O(l); \Delta H = -890 \text{ kJ/mol}$$

$$2C_2H_2(g) + 5O_2(g) \rightarrow 4CO_2(g) + 2H_2O(l); \Delta H = -2600 \text{ kJ/mol}$$

Which of the two fuels produces the larger quantity of energy per gram of fuel consumed?

5. Space rockets carry liquid hydrogen and liquid oxygen. They generate energy from the combustion:

$$2H_2(g) + O_2(g) \rightarrow 2H_2O(g); \Delta H = -480 \text{ kJ/mol of reaction}$$

Calculate the quantity of energy liberated when 1.00 g of steam is produced.

6. People are becoming more interested in the possibility of using hydrogen as a fuel.

$$2H_2(g) + O_2(g) \rightarrow 2H_2O(l); \Delta H = -570 \text{ kJ/mol of reaction}$$

Methane, $CH_4$, is the fuel we burn in North Sea gas.

$$CH_4(g) + 2O_2(g) \rightarrow CO_2(g) + 2H_2O(l); \Delta H = -890 \text{ kJ/mol}$$

Coal, if it were pure carbon, would burn according to the equation

$$C(s) + O_2(g) \rightarrow CO_2(g); \Delta H = -390 \text{ kJ/mol}$$

Which of the three fuels, hydrogen, methane or coal, provides the most energy per gram of fuel burned?

7. The molar heats of combustion of some alcohols are tabulated below.

| Alcohol | Formula | $\Delta H$ (kJ/mol) |
|---|---|---|
| Methanol | $CH_3OH$ | −720 |
| Ethanol | $CH_3CH_2OH$ | −1370 |
| Butanol | $CH_3(CH_2)_2CH_2OH$ | −2670 |
| Pentanol | $CH_3(CH_2)_3CH_2OH$ | −3300 |

(a) Plot a graph of $\Delta H$ (vertical axis) against the number of carbon atoms (horizontal axis).

(b) From your graph, estimate $\Delta H$ for
(i) propanol, $CH_3CH_2CH_2OH$
(ii) hexanol, $CH_3(CH_2)_4CH_2OH$.

8. A solution of sodium hydroxide was made by dissolving 80 g and making up to 1 $dm^3$. 50 $cm^3$ portions of this solution were placed in insulated plastic cups. The temperature of each was noted. A different volume of dilute hydrochloric acid was added to each cup. The temperature rise was noted, and the heat evolved was calculated. The results are shown in the table.

| Volume of sodium hydroxide solution ($cm^3$) | Volume of hydrochloric acid ($cm^3$) | Heat evolved (kJ) |
|---|---|---|
| 50 | 10 | 1.1 |
| 50 | 20 | 2.2 |
| 50 | 30 | 3.4 |
| 50 | 40 | 4.5 |
| 50 | 50 | 5.6 |
| 50 | 60 | 5.6 |
| 50 | 70 | 5.6 |

(a) Why does the heat change become constant after 50 $cm^3$ of acid have been added?

(b) What amount in moles of sodium hydroxide is present in each cup?

(c) Calculate the concentration of hydrochloric acid in mol/$dm^3$.

(d) What quantity of heat will be evolved if a solution containing 1 mole of sodium hydroxide is neutralised by hydrochloric acid?

(e) Estimate the quantity of heat that will be evolved if a solution containing 1 mole of sodium hydroxide is neutralised by sulphuric acid.

(f) Estimate the heat evolved if a solution containing 1 mole of $Ca(OH)_2$ is neutralised by hydrochloric acid.

9. 25.0 $cm^3$ of a solution of potassium hydroxide were poured into an insulated beaker. 2.00 mol/$dm^3$ nitric acid was added from a burette. The temperature of the solution was recorded at intervals. The results are tabulated overleaf.

| Volume of nitric acid added ($cm^3$) | Temperature (°C) |
| --- | --- |
| 0 | 18.0 |
| 5.0 | 20.5 |
| 10.0 | 23.0 |
| 15.0 | 25.5 |
| 20.0 | 27.9 |
| 25.0 | 30.3 |
| 30.0 | 32.9 |
| 35.0 | 31.6 |
| 40.0 | 30.5 |

(a) On graph paper, plot the temperature (vertical axis) against the volume of nitric acid added (horizontal axis).

(b) State the volume of 2.00 mol/$dm^3$ nitric acid which exactly neutralises 25.0 $cm^3$ of potassium hydroxide solution.

(c) Calculate the concentration of the potassium hydroxide solution.

# 8. The Speeds of Chemical Reactions

## Reaction Speeds

Steps can be taken to speed up a chemical reaction, or to slow it down. There are four factors which influence the speed of a chemical reaction. They are:

1. the particle size of the solid reactants
2. the concentrations of the reacting solutions
3. the temperature
4. the presence of a catalyst.

In most catalysed reactions, a very small quantity of catalyst is all that is required. The reaction does not go faster still if you add more catalyst.

## Graphs which show the extent of reaction plotted against time

The speed at which a substance in solution reacts is proportional to the concentration of the solution. As more and more of the reactant is used up, the reaction becomes slower and slower. The progress of the reaction has the form shown in Figure 8.1.

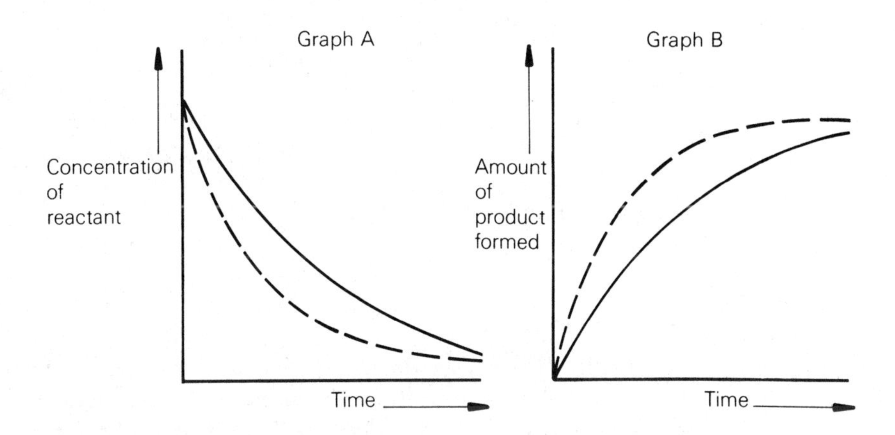


Figure 8.1 Graphs of extent of reaction against time

Graph A shows the way in which the concentration of a reactant decreases as the reaction takes place. Graph B shows how the amount of product increases as the time interval since the beginning of the reaction increases. For example, if you collect the oxygen formed in the decomposition of hydrogen peroxide in a syringe, and read off its volume at various intervals of time after the beginning of the reaction, the plot of the volume of gas formed against time will have the form shown in Graph B.

If the reaction is speeded up, the reactant will be used up in a shorter time (as shown by the broken line in A), and the product will be formed in a shorter time (as shown by the broken line in B). Methods of speeding up the reaction do not affect the amount of product formed. 34 g of hydrogen peroxide will give a maximum of 12.0 $dm^3$ of oxygen (at r.t.p.) whether the reaction occurs at room temperature or at 80 °C, whether the reaction occurs slowly over a period of weeks or quickly in the presence of a catalyst. You cannot create more product by altering the conditions.

**Example 1** Figure 8.2 shows a plot of the volume of oxygen produced from the decomposition of hydrogen peroxide against the time for which the reaction has been occurring. State whether A or B or C represents the fastest reaction.

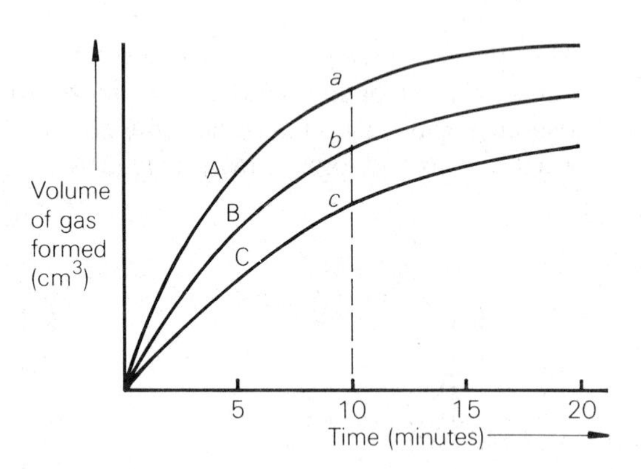


Figure 8.2 Plot of volume of oxygen against time

**Method:** Read upwards from a time of 10 minutes on the time axis. In reaction A, *a* $cm^3$ of gas have been formed; in reaction B *b* $cm^3$ of gas have been formed and in reaction C, *c* $cm^3$ of gas have been formed. You can see that volume *a* is more than volume *b*, which is more than *c*, so that reaction A is faster than B, which is faster than C.

**Example 2**

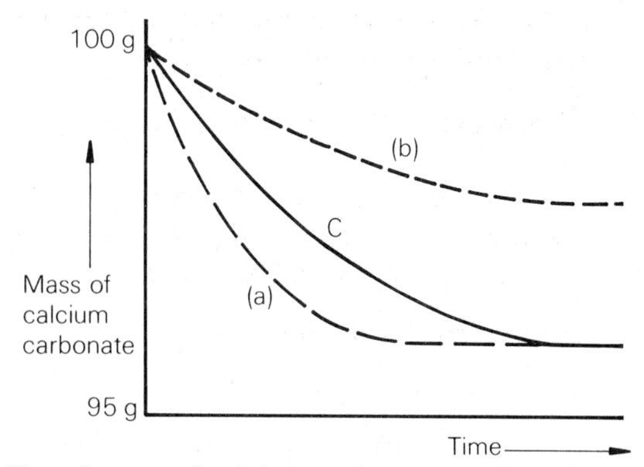


Figure 8.3 Plot of mass of calcium carbonate against time

Curve C in Figure 8.3 shows how the mass of calcium carbonate decreases with time when 100 g of marble chips are allowed to react with 50 cm$^3$ of 2 mol/dm$^3$ hydrochloric acid. On the graph, mark the plots you would obtain (a) by carrying out the reaction at 30 °C, (b) by using 50 cm$^3$ of 1 mol/dm$^3$ hydrochloric acid.

**Method:** At a higher temperature, the rate of reaction will be faster. Curve (a), therefore, lies to the left of curve C.

Using a lower concentration of acid, the rate will be slower, and plot (b), therefore, lies to the right of curve C. With 50 cm$^3$ of 1 mol/dm$^3$ acid (instead of 2 mol/dm$^3$) only half as much calcium carbonate will be able to react, and there will be only half the decrease in mass. Curve (b) levels off after a decrease of 2.5 g, instead of 5 g.

**Example 3** The volume of carbon dioxide formed by the reaction of an excess of calcium carbonate with 50 cm$^3$ of 0.1 mol/dm$^3$ hydrochloric acid is plotted against time, in Figure 8.4.

(a) What volume of gas has been formed after 5 minutes?

(b) How long does the reaction take to produce 50 cm$^3$ of gas?

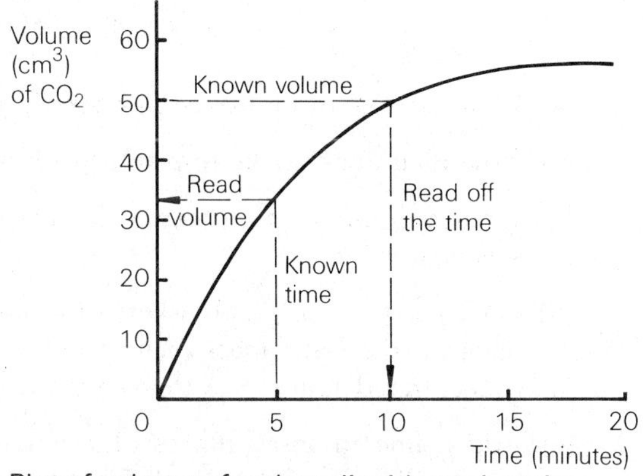


Figure 8.4 Plot of volume of carbon dioxide against time

**Method:** (a) Draw a vertical line from the stated time of 5 minutes. Where it cuts the graph, draw a horizontal line across to the volume axis, and read off the volume.

**Answer:** 32 $cm^3$ of carbon dioxide.

(b) Draw a horizontal line across from the stated volume, 50 $cm^3$, to cut the graph. Where it cuts the graph, drop a vertical line down to the time axis, and read off the time.

**Answer:** 10 minutes.

## Problems on Reaction Speeds

1. The graph shows the total volume of hydrogen produced in the reaction of magnesium ribbon with an excess of dilute hydrochloric acid over a period of time.

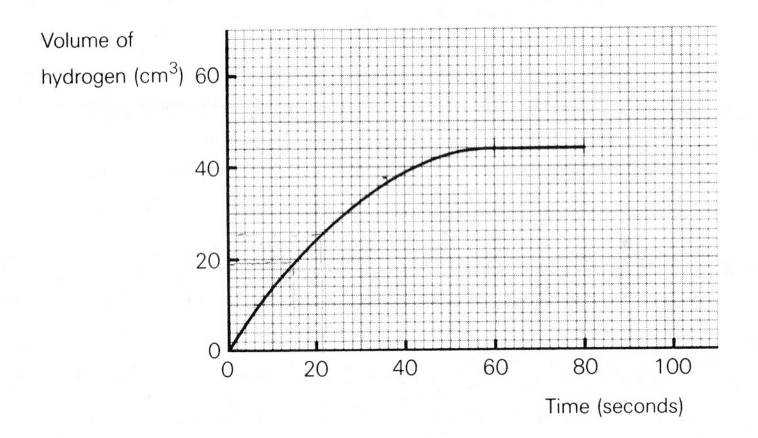


(a) What volume of hydrogen has been produced after 15 seconds?

(b) How long does it take to produce 25 $cm^3$ of hydrogen?

(c) What volume of hydrogen will have been produced after 100 seconds?

(d) On a copy of the graph, sketch the results you would expect to obtain if the same mass of magnesium was treated with a more concentrated solution of the same acid.

(e) Add a line to show the results you would expect if you used magnesium powder instead of magnesium ribbon.

2. Calcium carbonate was placed in a flask on a balance, and dilute hydrochloric acid was added. The mass of the flask and its contents was recorded every minute. A plot of the results is shown below.

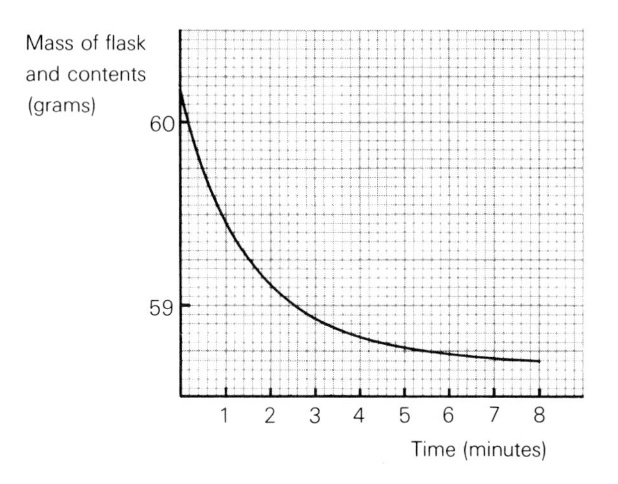


At which of the following times was the reaction fastest?

**A** 1 minute

**B** 2 minutes

**C** 3 minutes

**D** 4 minutes

**E** 5 minutes

3. An excess of acid was added to some marble chips. The volume of carbon dioxide was measured every minute, and the results were plotted.

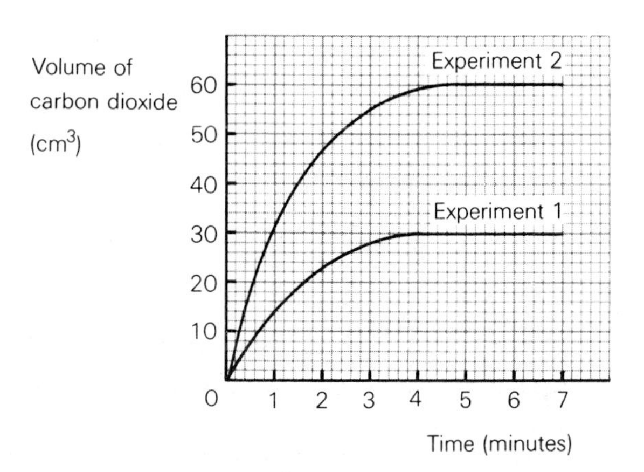

The results show that

**A** twice as much acid was used in Experiment 2.

**B** Experiment 2 was done at a higher temperature.

**C** twice as much marble was used in Experiment 2.

**D** smaller chips of marble were used in Experiment 2.

**E** Experiment 2 was done using a more concentrated acid.

4. In two experiments, two equal masses of magnesium ribbon were allowed to react with two 50 $cm^3$ portions of dilute hydrochloric acid. The gas evolved was collected and measured. The volume of gas was plotted against the time since the start of the reaction. In Experiment 1, the temperature was 20 °C; in Experiment 2, the temperature was 50 °C. Which of the graphs shown here best shows the results of the two experiments?

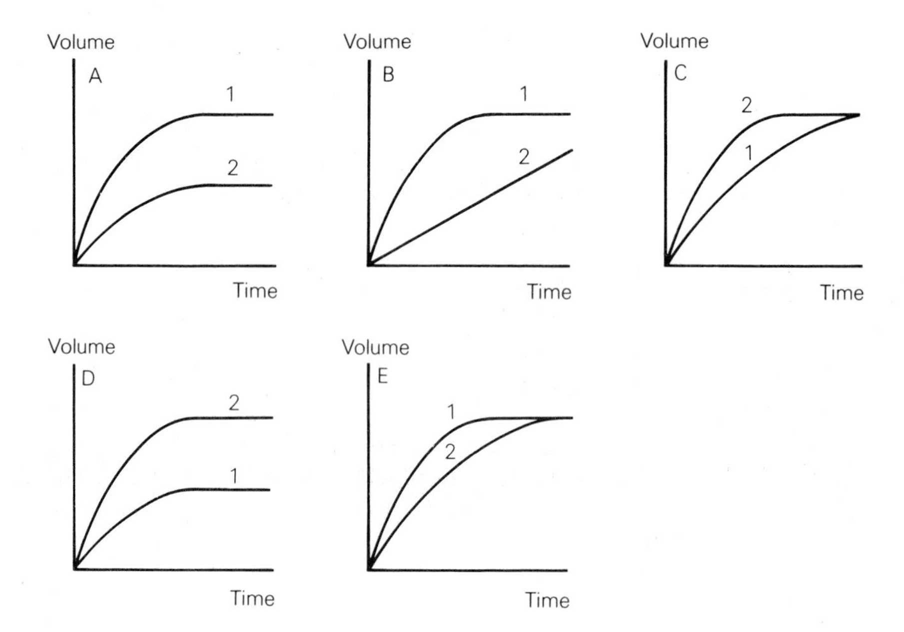


5. An experiment was carried out to find the speed at which marble reacts with hydrochloric acid. A conical flask containing 25.0 $cm^3$ of 1.00 $mol/dm^3$ hydrochloric acid was placed on a top-loading balance. 10.0 g of marble chips were added, and the neck of the flask was stoppered with cotton wool. The mass of the flask and its contents was measured at the start of the reaction and at 2 minute intervals for 12 minutes. The results are shown in the table.

| Time (minutes) | 0 | 2 | 4 | 6 | 8 | 10 | 12 |
|---|---|---|---|---|---|---|---|
| Mass of flask and contents (g) | 102.03 | 101.74 | 101.61 | 101.54 | 101.50 | 101.48 | 101.48 |

(a) Plot these results on graph paper. Use a scale 1 cm = 1 minute along the horizontal axis and 1 cm = 0.1 g along the vertical axis.

(b) What mass of carbon dioxide is formed?

(c) How long does the reaction take to be completed?

(d) How long does it take for half the total mass of carbon dioxide to be given off?

(e) Why is the time in (d) different from half the time in (c)?

(f) On your graph, draw the curve you would expect to obtain if you used 25.0 $cm^3$ of 0.50 $mol/dm^3$ hydrochloric acid in a similar experiment.

(g) Add a curve showing the result you would expect to obtain if you crushed the 10.0 g of marble chips before adding them to the acid.

6. Hydrogen peroxide, $H_2O_2$, decomposes to give oxygen, $O_2$, and water, $H_2O$. An experiment was performed in which a catalyst was added to speed up the decomposition and the volume of oxygen was measured at various times after the start of the reaction. The following results were obtained.

| Time (minutes) | 1.0 | 2.0 | 3.0 | 4.0 | 5.0 | 6.0 | 7.0 |
|---|---|---|---|---|---|---|---|
| Volume ($cm^3$) | 11.0 | 18.0 | 21.2 | 23.4 | 25.2 | 26.6 | 27.8 |

(a) Write a balanced chemical equation for the reaction.

(b) Plot the volume of oxygen used against time.

(c) From your graph, say what volume of oxygen would be formed in 2.5 minutes.

(d) How long would it take to produce 24.0 $cm^3$ of oxygen?

7. When sodium thiosulphate reacts with acid, a yellow suspension of sulphur forms.

$$Na_2S_2O_3(aq) + 2HCl(aq) \rightarrow S(s) + 2NaCl(aq) + SO_2(g) + H_2O(l)$$

The effect of temperature on this reaction was studied. A 25.0 $cm^3$ portion of a 0.02 $mol/dm^3$ solution of sodium thiosulphate was run into a conical flask. A cross was drawn on a piece of paper underneath the flask. An excess of hydrochloric acid was added, and the temperature was noted. The time taken for the cross to disappear was noted. The experiment was repeated at different temperatures. The results are shown in the table.

| Temperature (°C) | 19 | 29 | 40 | 49 | 60 |
|---|---|---|---|---|---|
| Time (seconds) | 340 | 250 | 155 | 90 | 55 |

(a) Plot a graph of time (vertical axis) against temperature (horizontal axis).

(b) From your graph estimate the reaction time at (i) 25 °C (ii) 54 °C.

(c) What effect does an increase in temperature have on the speed of the reaction?

(d) What effect would an increase in the concentration of the sodium thiosulphate solution have on the speed of the reaction?

(e) What would the reaction time be if 25.0 $cm^3$ of 0.20 $mol/dm^3$ solution were used at 35 °C?

# 9. Solubility

## Solubility

**Solubility** is defined as the mass in grams of solute required to saturate 100 g of solvent at a stated temperature. A **saturated solution** is one that contains the maximum amount of solute that can be dissolved at that temperature. When values of solubility are plotted against temperature, the graph obtained is called a **solubility curve**. Figure 9.1 shows solubility curves for a number of substances over temperatures of 0 °C to 100 °C.

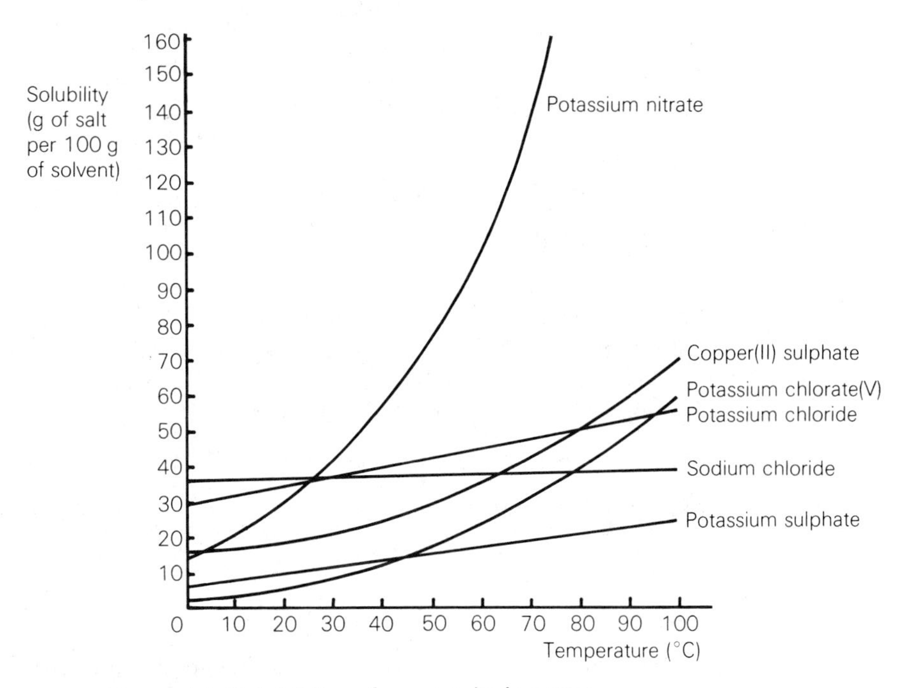


Figure 9.1 Solubilities of some salts in water

**Example 1** If 8 g of potassium chloride saturate 20 g of water at 40 °C, what is the solubility of potassium chloride?

**Method:** Problems of this kind are a simple ratio type of calculation.

If 20 g water are saturated by 8 g potassium chloride,

100 g water are saturated by $\frac{100 \times 8}{20} = 40$ g potassium chloride.

**Answer:** Solubility of potassium chloride at 40 °C = 40 g per 100 g of water.

**Example 2** 100 g of water at 80 °C are saturated with potassium nitrate. What mass of potassium nitrate will crystallise out if the solution is cooled to 20 °C?

**Method:** Look at the solubility curves in Figure 9.1. Draw a vertical line from 80 °C to cut the solubility curve of potassium nitrate. On the solubility axis, read off what mass of potassium nitrate is dissolved. You can see that 170 g of potassium nitrate are dissolved in 100 g of water at 80 °C. Draw a vertical line from 20 °C to cut the solubility curve. You can read off a value of 35 g per 100 g of water. On cooling from 80 °C to 20 °C, the solubility drops from 170 to 35 g per 100 g of water. The difference, 135 g of potassium nitrate, comes out of solution.

**Answer:** 135 g of potassium nitrate come out of solution on cooling 100 g of water saturated with this salt from 80 °C to 20 °C.

## Problems on Solubility

1. Calculate the solubility of sodium chloride in water at 20 °C from the following results. They were obtained by evaporating a known mass of a saturated solution of sodium chloride to dryness and then finding out what mass of sodium chloride was left.

   Mass of evaporating basin = 21.45 g

   Mass of evaporating basin + saturated solution = 38.45 g

   Mass of sodium chloride = 25.95 g

2. Values of the solubility of potassium nitrate are given below.

| Temperature (°C) | Solubility (g/100 g water) |
|---|---|
| 20 | 32 |
| 50 | 86 |

   When 93 g of a saturated solution of potassium nitrate are cooled from 50 °C to 20 °C, what mass of crystals should form?

3. Values of the solubility of potassium sulphate are given opposite.

| Temperature (°C) | Solubility (g/100 g water) |
|---|---|
| 20 | 12 |
| 80 | 20 |

What will happen when 60 g of a saturated solution of potassium sulphate are cooled from 80 °C to 20 °C?

4. 20.0 g of potassium chloride were placed in a Pyrex beaker and 40.0 $cm^3$ of water were added. The beaker was heated until all the potassium chloride had dissolved and then allowed to cool. When crystals first appeared, the temperature was noted. An extra 5 $cm^3$ of water were added, and the experiment was repeated. The results of experiments are shown below.

| Experiment | Volume of water ($cm^3$) | Temperature at which crystals formed (°C) | Solubility (g/100 $cm^3$ water) |
|---|---|---|---|
| 1 | 40 | 77 | . . . |
| 2 | 45 | 56 | 44.5 |
| 3 | 50 | 40 | . . . |
| 4 | 55 | 26 | 36.3 |
| 5 | 60 | 15 | . . . |
| 6 | 65 | 8 | 30.8 |

(a) Calculate the values of solubility in g/100 $cm^3$ water which are missing from the table.

(b) Plot the values of solubility (vertical axis) against temperature (horizontal axis).

(c) What is the effect of temperature on the solubility of potassium chloride in water?

(d) What is the solubility of potassium chloride at 50 °C?

5. Plot solubility curves for potassium chloride and potassium chlorate (V) from the following data.

| Temperature (°C) | 0 | 20 | 40 | 60 | 80 | 100 |
|---|---|---|---|---|---|---|
| Potassium chloride (g/100 g water) | 28.0 | 33.0 | 38.0 | 43.0 | 48.0 | 53.0 |
| Potassium chlorate (V) (g/100 g water) | 4.0 | 7.5 | 14.0 | 25.0 | 41.0 | 59.0 |

If 200 g of a saturated solution of the two salts was cooled from 90 °C to 15 °C, what mass of crystals would be obtained?

6. Some solubility curves are shown in Figure 9.2

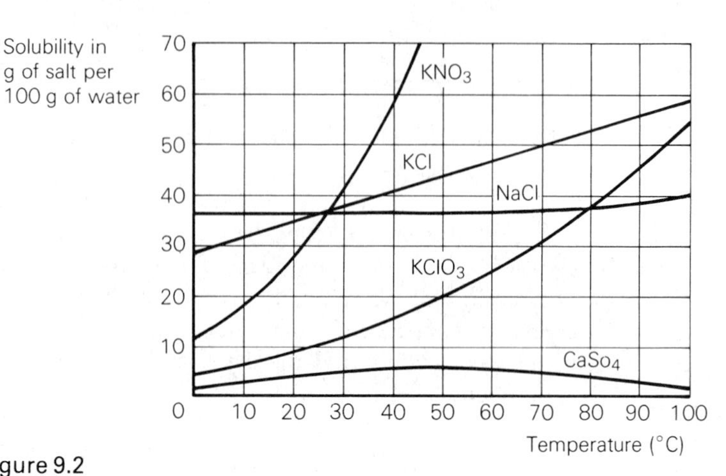


Figure 9.2

(a) Which is the most soluble salt at 20 °C?

(b) What is the maximum mass of potassium chloride that will dissolve in 50 g of water at 70 °C?

(c) If 50 g of potassium chlorate (V) ($KClO_3$) were dissolved in 100 g of water at 100 °C and the solution was cooled to 5O °C, what mass of crystals would be deposited?

# 10. Radioactivity

## Radioactivity

A large number of elements are **radioactive**: their nuclei are unstable and split to form two new nuclei. This type of reaction is called a **nuclear reaction**. It is quite different from a chemical reaction in which the atoms stay the same and only the bonding changes. Sometimes, protons, neutrons and electrons fly out when the original nucleus divides. The particles and energy given out are called **radioactivity**, and the breaking-up process is called **radioactive decay**.

*There are three types of radiation*: $\alpha$-rays, $\beta$-rays and $\gamma$-rays. $\beta$-rays and $\gamma$-rays are penetrating rays, and can be detected by means of an instrument called a Geiger-Müller counter. Each time a $\beta$-ray or a $\gamma$-ray is emitted, it ionises the gas inside the Geiger-Müller counter, and a pulse of electricity flows through the counter. The pulses of electricity can be fed into a loudspeaker to produce a click or counted on an electronic counter.

The results of measurements of radioactive decay always have the form shown in Figure 10.1.

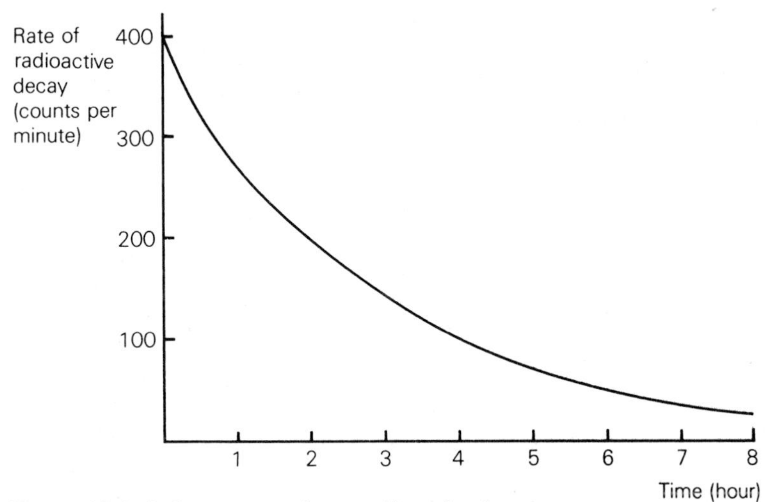


Figure 10.1 A decay curve for a radioactive isotope

You can see that

the time taken for the counts to fall from 400 to 200 c.p.m. is 2 hours, and
the time taken for the counts to fall from 200 to 100 c.p.m. is 2 hours, and
the time taken for the counts to fall from 100 to 50 c.p.m. is also 2 hours.

The time taken for the rate of radioactive decay to fall to half its value is the same, no matter what the original rate of decay is. This time is called the **half-life**. Radioactive elements differ enormously in their half-lives, from a fraction of a second to millions of years.

You often see elements referred to as 'radioactive isotopes'. Some elements have atoms of more than one kind. The different kinds of atoms are called **isotopes**. They differ in the number of neutrons in the nucleus. Thus, they have the same atomic number (the number of protons) and different mass numbers (the number of protons plus the number of neutrons). They are represented as

$$^{\text{Mass number}}_{\text{Atomic number}}\text{Symbol}$$

For example, the isotopes of carbon are $^{12}_{6}C$, $^{13}_{6}C$ and $^{14}_{6}C$. They are often referred to as carbon-12, carbon-13 and carbon-14.

## Examples of Problems on Radioactivity: Type 1

Problems on radioactivity are very simple. You may be told the half-life of a radioactive isotope and asked to work out how the count rate will decrease with time. You may, given the half-life, be asked to calculate how the mass of a radioactive isotope decreases with time. Another kind of problem is to work out the half-life of a radioactive decay from a plot of counts per minute against time.

**Example 1** The half-life of bromine-82 is 36 hours. A sample of potassium bromide solution containing bromine-82 is put into a Geiger-Müller counter. The count rate is 160 c.p.m. What will be the value of the count rate 72 hours later?

**Method:**
Count rate = 160 c.p.m.
Half-life = 36 hours.
The count rate will have fallen to half its value in 36 hours.
Count rate after 36 hours = 80 c.p.m.
In a further 36 hours, the count rate will fall to half its value, i.e. from 80 to 40 c.p.m.
Count rate after 72 hours = 40 c.p.m.

**Answer:** Count rate after 72 hours = 40 c.p.m.

**Example 2** Caesium-137 has a half-life of 30 years. A sample of caesium chloride contains 2.00 g of caesium-137. What will be the mass of caesium-137 left after 90 years?

**Method:**
Mass of radioactive isotope = 2.00 g
Half-life = 30 years
Time span = 90 years = 3 half-lives
Mass of radioactive isotope left after 3 half-lives = $2 \times \frac{1}{2} \times \frac{1}{2} \times \frac{1}{2}$
$= 0.25$ g.

**Answer:** 0.25 g of caesium-137 will remain after 90 years.

**Example 3** A solution of a radioisotope was placed in a Geiger-Müller counter, and the count rate was measured at intervals over a number of days. The results are shown in Table 10.1. Choosing a suitable scale, plot the rate of radioactive decay against time. From your graph, find the half-life of the nuclear reaction.

Table 10.1

| Time (days) | Decay rate (c.p.m.) |
|---|---|
| 0 | 1000 |
| 1 | 820 |
| 2 | 660 |
| 3 | 520 |
| 4 | 420 |
| 5 | 340 |
| 6 | 260 |
| 8 | 170 |

**Method:** The first thing you have to do is choose a scale. Along the horizontal axis, 1 cm = 1 day will spread the results out nicely. Along the vertical axis, 1 cm = 100 c.p.m. will be suitable.

The plot of the results is shown in Figure 10.2.

At zero time, the decay rate is 1000 c.p.m. The half-life of the radioactive decay is the time taken for the c.p.m. to fall to half this value, i.e. to 500 c.p.m. Draw a horizontal line from 500 c.p.m. across to intersect the graph. Drop a vertical line to the $x$ axis to find the time which has passed. The time is 3.2 days.

Now find the time taken for the count rate to fall from 500 to 250 c.p.m. Draw a horizontal line across to cut the graph. Drop a vertical line to the $x$ axis. Read off the time. The time is 6.4 days. To fall from 500 c.p.m. to 250 c.p.m. takes from 6.4 to 3.2 days, that is 3.2 days.

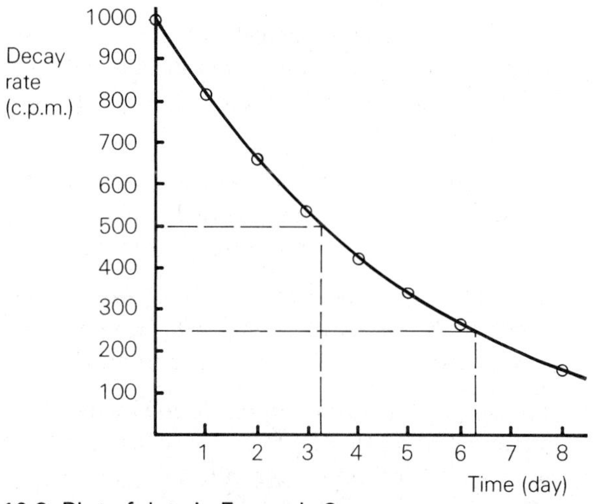


Figure 10.2 Plot of data in Example 3

The graph will tell you the time taken for the count rate to fall from 800 c.p.m. to 400 c.p.m., from 600 to 300 c.p.m. and so on. Whichever part of the graph you use will give the same answer for the half-life.

## Examples of Problems on Radioactivity: Type 2

There is another type of numerical problem on radioactivity, in which you are asked to fill in the missing mass numbers and atomic numbers in the equation for a nuclear reaction.

**Example 1** The radioactive isotope of sodium decays according to the equation

$$^{24}_{11}\text{Na} \rightarrow {}^{0}_{-1}\text{e} + {}^{m}_{n}\text{X}$$

Fill in the mass number, $m$, and the atomic number, $n$, of the element, X.

**Method:** The sum of the mass numbers on the left-hand side of the equation equals the sum of the mass numbers on the right-hand side of the equation. Therefore,

$$24 = 0 + m$$
$$m = 24$$

Sum of atomic numbers on LHS = Sum of atomic numbers on RHS. Therefore,

$$11 = -1 + n$$
$$n = 12$$

**Answer:** The isotope formed is $^{24}_{12}\text{X}$.

## Problems on Radioactivity

## Section 1

1. Californium is a radioactive element with a half-life of 44 minutes. If a sample of californium is giving a count rate of 500 counts per minute, what will be the count rate after (a) 44 minutes, (b) 88 minutes, (c) 132 minutes?
2. A scientist succeeds in isolating 0.4 g of the radioactive isotope, einsteinium. It has a half-life of 25 minutes. What mass of einsteinium will remain after 50 minutes, and what mass after 100 minutes?
3. A sample of nobelium, a radioactive isotope with a half-life of 3 seconds, has a count rate of 160 counts per minute. How long will it take to decrease to 20 counts per minute?
4. Plutonium has a half-life of 24 360 years. A sample of plutonium is decaying at a rate of 200 counts per minute. How long will it take to decrease to 50 counts per minute?

5. In the following equations for nuclear reactions, there are some spaces. Supply the missing mass numbers and atomic numbers.
   (a) ${}^{210}_{82}Pb \rightarrow \quad Bi + {}^{0}_{-1}e$
   (b) ${}^{210}_{84}Po \rightarrow \quad Pb + {}^{4}_{2}He$
   (c) ${}^{27}_{13}Al + {}^{1}_{1}H \rightarrow {}^{4}_{2}He + \quad Mg$
   (d) $\quad X + {}^{4}_{2}He \rightarrow {}^{17}_{9}F + {}^{1}_{0}n$
   (e) ${}^{11}_{5}B + {}^{1}_{1}H \rightarrow \quad X + {}^{1}_{0}n$
   (f) ${}^{24}_{12}Mg + {}^{4}_{2}He \rightarrow {}_{14}Si + {}^{1}Y$

## Section 2

1. The half-life of carbon-14 is 5700 years. A sample of carbon-14 gives a count rate of 84 counts per minute on a Geiger-Müller counter. How long would it take for the count rate to drop to 21 counts per minute?

2. Uranium-239 has a half-life of 24 minutes. A solution of uranium-239 nitrate gives a count rate of 8000 c.p.m. on a Geiger-Müller counter. What will be the count rate after (a) 48 minutes, and (b) 2 hours?

3. An enriched form of a radioactive isotope was obtained in solution. The solution was placed in a Geiger-Müller counter, and the radioactive count was measured at regular time intervals. The results are given in Table 10.2.

Table 10.2

| Time in minutes | Rate of decay (counts per second) |
|---|---|
| 0 | 600 |
| 30 | 225 |
| 60 | 81 |
| 90 | 30 |
| 120 | 11 |

   (a) Plot a graph of the count rate against time.
   (b) Deduce from your graph the count rate at 50 minutes and at 100 minutes.
   (c) After what length of time was the count rate (i) 200 c.p.s. and (ii) 100 c.p.s.?
   (d) Calculate the half-life of the radioactive element.
   (e) What would the count rate be at 144 minutes?
   (f) If a solution of half the concentration had been used, what would the count rate have been after 50 minutes and after 100 minutes?
   (g) What would the half-life have been, measured in this solution?

4. A sample of a radioactive isotope was put into a Geiger-Müller counter, and the count rate was measured at various times. The results are shown in Table 10.3 overleaf.

Table 10.3

| Time (days) | Count rate (counts per minute) |
|---|---|
| 0 | 520 |
| 50 | 410 |
| 100 | 320 |
| 200 | 195 |
| 300 | 115 |
| 400 | 70 |

Plot a graph of count rate against time. Use the horizontal axis for time.

(a) Use the graph to find the count rate after 150 days and after 250 days.

(b) After what length of time was the count rate (i) 300 c.p.m. (ii) 150 c.p.m.?

(c) What is the half-life of the radioactive element?

5. A radioactive isotope, *A*, has a half-life of 30 minutes. The initial rate of decay of *A* is 320 counts per minute.

(a) Draw a graph of the decay *A* over the first three hours.

(b) If the original sample of *A* had mass 64 g, what mass of *A* remains after three hours?

# Section 3

1. Strontium-90 is a radioactive isotope which is produced in nuclear reactors. It has a half-life of 28 years. Answer the following questions about a 100 g sample of strontium-90.

(a) How long will it take for 75 g of the sample to decay?

(b) How many grams will be left after 4 half-lives?

(c) How many grams will be left after 168 years?

(d) How many half-lives must elapse before less than 1 g remains?

2. State the number of (i) protons (ii) electrons and (iii) neutrons in the following atoms.

(a) ${}^{23}_{11}Na$ (b) ${}^{16}_{8}O$ (c) ${}^{57}_{25}Mn$ (d) ${}^{60}_{27}Co$ (e) ${}^{208}_{82}Pb$

3. The oldest known rocks on Earth are in Greenland. They have been dated by measuring the fraction of potassium-40 present in the potassium compounds in the rocks. It was found that

$$\frac{\text{K-40 originally present in the rocks}}{\text{K-40 now present in rocks}} = 8$$

If the half-life of K-40 is $1.3 \times 10^9$ years, how old are the rocks?

4. Copy the table, and fill in the missing information.

| Element | Number of protons | Number of electrons | Number of neutrons | Atomic number | Mass number |
|---|---|---|---|---|---|
| A | 35 | ... | 80 | ... | ... |
| B | ... | 17 | 18 | ... | ... |
| C | ... | ... | 60 | ... | 107 |
| D | 90 | ... | ... | ... | 232 |
| E | ... | ... | 33 | ... | 60 |

5. Supply the missing mass number and atomic number of E.

(a) $^{144}_{60}Nd \rightarrow {}^{140}_{58}Ce + E$

(b) $E \rightarrow {}^{90}_{38}Sr + {}^{0}_{-1}e$

(c) $^{150}_{64}Gd \rightarrow E + {}^{4}_{2}He$

(d) $^{234}_{91}Pa \rightarrow E + {}^{0}_{-1}e$

6. The following equations show nuclear reactions which happen when atoms are bombarded with small particles. (i) Supply the missing mass numbers and atomic numbers. (ii) Can you identify the bombarding particles?

(a) $^{27}_{13}Al + X \rightarrow {}^{30}_{15}P + {}^{1}_{0}n$

(b) $^{96}_{42}Mo + X \rightarrow {}^{97}_{43}Tc + {}^{1}_{0}n$

(c) $^{54}_{25}Mn + X \rightarrow {}^{54}_{24}Cr$

(d) $^{54}_{26}Fe + X \rightarrow {}^{54}_{25}Mn + {}^{1}_{1}p$

# Answers to Problems

## Chapter 1

### Practice with Equations

1. (a) $H_2(g) + CuO(s) \rightarrow Cu(s) + H_2O(g)$
   (b) $C(s) + CO_2(g) \rightarrow 2CO(g)$
   (c) $C(s) + O_2(g) \rightarrow CO_2(g)$
   (d) $Mg(s) + H_2SO_4(aq) \rightarrow H_2(g) + MgSO_4(aq)$
   (e) $Cu(s) + Cl_2(g) \rightarrow CuCl_2(s)$
2. (a) $Ca(s) + 2H_2O(l) \rightarrow H_2(g) + Ca(OH)_2(aq)$
   (b) $2Cu(s) + O_2(g) \rightarrow 2CuO(s)$
   (c) $4Na(s) + O_2(g) \rightarrow 2Na_2O(s)$
   (d) $Fe(s) + 2HCl(aq) \rightarrow FeCl_2(aq) + H_2(g)$
   (e) $2Fe(s) + 3Cl_2(g) \rightarrow 2FeCl_3(s)$
3. (a) $Na_2O(s) + H_2O(l) \rightarrow 2NaOH(aq)$
   (b) $2KClO_3(s) \rightarrow 2KCl(s) + 3O_2(g)$
   (c) $2H_2O_2(aq) \rightarrow 2H_2O(l) + O_2(g)$
   (d) $3Fe(s) + 2O_2(g) \rightarrow Fe_3O_4(s)$
   (e) $3Mg(s) + N_2(g) \rightarrow Mg_3N_2(s)$
   (f) $4NH_3(g) + 3O_2(g) \rightarrow 2N_2(g) + 6H_2O(g)$
   (g) $3Fe(s) + 4H_2O(g) \rightarrow Fe_3O_4(s) + 4H_2(g)$
   (h) $2H_2S(g) + 3O_2(g) \rightarrow 2H_2O(g) + 2SO_2(g)$
   (i) $2H_2S(g) + SO_2(g) \rightarrow 2H_2O(l) + 3S(s)$

## Chapter 2

### Problems on Relative Molecular Mass

| | | |
|---|---|---|
| 64 | 40 | 101 |
| 84 | 278 | 95 |
| 148 | 99 | 161 |
| 98 | 63 | 246 |
| 136 | 685 | 142 |
| 106 | 74 | 123.5 |
| 159.5 | 162 | 249.5 |
| 400 | 286 | 278 |

### Problems on Percentage Composition

#### Section 1

1. Mg = 60% O = 40%
2. Ca = 40% C = 12%
   O = 48%
3. K = 39% H = 1%
   C = 12% O = 48%
4. (a) N = 46.7% O = 53.3%
   (b) H = 5% F = 95%
   (c) Be = 36% O = 64%
   (d) Li = 46.7% O = 53.3%
5. (a) C = 80% H = 20%
   (b) Na = 57.5% O = 40%
   H = 2.5%
   (c) S = 40% O = 60%
   (d) C = 90% H = 10%
6. (a) C = 84% H = 16%
   (b) Mg = 72% N = 28%
   (c) Na = 15.3% I = 84.7%
   (d) Ca = 20% Br = 80%

#### Section 2

1. (a) C = 85.7% H = 14.3%
   (b) N = 35% H = 5%
   O = 60%
   (c) Fe = 62.2% O = 35.6%
   H = 2.2%
   (d) C = 26.7% H = 2.2%
   O = 71.1%
2. (a) Fe = 28% S = 24%
   O = 48%
   (b) 40.5% (c) 67.5%
   (d) 46.7%
3. (a) C = 60% H = 13%
   O = 27%
   (b) C = 40.0% H = 6.7%
   O = 53.3%
   (c) C = 40.0% H = 6.7%
   O = 53.3%
   (d) Al = 36% S = 64%

#### Section 3

1. 34 000
2. 1.1 kg

# Chapter 3

## Problems on the Mole

### Section 1

1. (a) 23 g (b) 24 g (c) 207 g
2. (a) 13.7 g (b) 5.2 g (c) 11.9 g
3. (a) 508 g (b) 216 g (c) 54 g
   (d) 402 g
4. (a) 27 g (b) 8 g (c) 6 g
   (d) 10 g (e) 5 g
5. (a) 2.0 (b) 0.05 (c) 0.75
   (d) 0.50 (e) 0.20 mol
6. (a) 44 g (b) 98 g (c) 36.5 g
   (d) 40 g
7. (a) 58.5 g (b) 28 g (c) 508 g
   (d) 265 g (e) 13.6 g
8. (a) 111 g; 123.5 g; 171 g; 85 g
   (b) 27.75 g; 30.87 g; 42.75 g;
   21.25 g

### Section 2

1. (a) 26 g (b) 8 g (c) 4 g
   (d) 6 g (e) 4 g (f) 8 g
2. (a) 2.0 (b) 2.0 (c) 0.25
   (d) 0.10 (e) 0.25 mol
3. (a) 23 g (b) 7 g (c) 14 g
   (d) 8 g (e) 16 g (f) 32 g
4. (a) 1.0 (b) 2.0 (c) 0.33
   (d) 3.0 (e) 0.5 (f) 0.125 mol
5. (a) 2070 g (b) 10.6 g
   (c) 25.4 g (d) 20.0 g
   (e) 10.0 g (f) 40.0 g
   (g) 42.0 g (h) 13.0 g
   (i) 35.5 g (j) 2.00 g
6. (a) 1.00 (b) 0.25
   (c) 0.50 (d) 0.20
   (e) 0.20 (f) 3.0
   (g) 0.10 (h) 2.0
7. (a) $6 \times 10^{23}$ (b) $6 \times 10^{23}$
   (c) $6 \times 10^{22}$ (d) $3.6 \times 10^{24}$
   (e) $1.2 \times 10^{24}$ (f) $6 \times 10^{22}$
   (g) $1.5 \times 10^{22}$ (h) $1.2 \times 10^{24}$
8. (a) 65 g (b) 0.065 g
9. (a) 9.0 g (b) 0.027 g
10. (a) 12 g (b) 0.040 g
11. (a) 20.0 g (b) 12.0 g
    16.25 g (d) 115 g

(b) 0.4 mol
l (d) 0.2 mol

79

3. $1.0 \times 10^{23}$
4. 55.5 mol
5. 6.3 mol
6. $3 \times 10^{-23}$ g
7. 2.92 mol

## Problems on Reacting Masses of Solids

### Section 1

1. 40 g
2. 10 g
3. 44 g
4. 14.7 g
5. 32 g
6. 4 g
7. 50 g
8. (a) 63.5 g (b) 12.7 g
9. 4.4 g
10. 1 g
11. 12 g
12. 2.8 g

### Section 2

1. 2.40 g
2. 23.2 g
3. 1 g
4. 22.3 g
5. 6.6 g
6. 8.0 g
7. 71 g
8. 27 g
9. 10.6 g
10. 170 g
11. C
12. D
13. B

### Section 3

1. 1250 tonne
2. 0.05 g
3. 3.06 kg
4. 0.26 g
5. (a) $2O_2$, $2H_2O$ (b) 2.25 kg
6. (a) $2Al(OH)_3$, $3H_2SO_4$, $3H_2O$
   (b) (i) 0.46 kg (ii) 0.86 kg
7. 25.0 kg
8. Loss of 83 p
9. (a) 72 g
10. (a) $3O_2 \rightarrow 2O_3$ (b) 64 g
11. 19 kg/year
12. (a) 1000 tonne (b) 1375 tonne
13. Yes, 1 mol C (12 g) combines with 4 mol $F^-$ (4 × 19 g)

14. 7.1 tonne
15. (a) 0.01 mol (b) 0.02 mol
    (c) 2 mol
    (d) $Zn(s) + 2Ag^{+}(aq) \rightarrow Zn^{2+}(aq) + 2Ag(s)$
16. 127
17. (a) 0.56 g (b) $8.0 \times 10^{-2}$ mol
18. (a) 4.05 tonne (b) 8.33 tonne
19. (a) (i) 1060 tonne (ii) 1030 tonne
    (b) natural limestone; manufactured ammonia
    (c) Ammonium sulphate is a fertiliser.

### Section 3

1. $FeCl_3$
2. Br
3. 6
4. A
5. D
6. $MO_2$

# Chapter 4

## Problems on Formulae

### Section 1

1. $Na_2O$
2. $Mg_3N_2$
3. $Fe_3O_4$
4. $HgBr_2$
5. $Al_2O_3$
6. $BaCl_2{\cdot}2H_2O$
7. $PbO_2$
8. (a) $C_7H_{16}$ (b) $Mg_3N_2$
   (c) $Al_2S_3$ (d) $CaBr_2$
   (e) $Cr_2S_3$

### Section 2

1. (a) $SO_2$ (b) $SO_3$
   (c) NO (d) $NO_2$
   (e) $CH_4$ (f) $CH_2$
2. (a) $P_2O_3$ (b) $NH_3$
   (c) $Pb_3O_4$ (d) $SiO_2$
   (e) $MnO_2$ (f) $N_2O_5$
   (g) $CrCl_3$
3. $A = C_2F_4$ $B = C_4H_8O_2$
   $C = C_2H_6$ $D = C_6H_6$
   $E = C_3H_6$ $F = C_2H_6O_2$
   $G = C_2H_4Cl_2$ $H = C_6H_3N_3O_6$
4. (a) $Na_2O$ (b) $Pb_3O_4$
   (c) $NO_2$ (d) $Cu_2O$
   (e) $FeCl_2$ (f) $FeCl_3$
5. (a) $CO_2$ (b) $PbO_2$
   (c) $CuCl_2$ (d) MgO
   (e) $Mg_3N_2$ (f) $AlBr_3$
6. (a) $MgSO_4{\cdot}7H_2O$
   (b) $CuSO_4{\cdot}5H_2O$
   (c) $Cr(NO_3)_3{\cdot}9H_2O$
7. (a) $MgSO_4$ (b) $N_2H_4O_3$
   (c) $C_3H_8O$ (d) $CH_2O$

# Chapter 5

## Problems on Reacting Volumes of Gases

### Section 1

1. 24 $dm^3$
2. 12 $dm^3$
3. 2.4 $dm^3$
4. 2.4 $dm^3$
5. 250 $cm^3$ $O_2$; 125 $cm^3$ $CO_2$
6. 125 $cm^3$ $O_2$

### Section 2

1. 20.0 $dm^3$ $O_2$; 20.0 $dm^3$ $CO_2$
2. 25 g; 6.0 $dm^3$
3. 3250 g; 1200 $dm^3$
4. 480 $cm^3$; 4.14 g
5. 120 $cm^3$
6. 600 $cm^3$
7. 0.41 g
8. 61.5 $cm^3$
9. 120 $dm^3$ $O_2$; 72.0 $dm^3$ $CO_2$
10. 600 $cm^3$; 1200 $cm^3$

### Section 3

1. (a) $2C_2H_6(g) + 7O_2(g) \rightarrow 4CO_2(g) + 6H_2O(g)$
   (b) 30 $cm^3$ (c) 40%
2. C
3. C
4. $CH_4$
5. 20 $cm^3$ ethane + 10 $cm^3$ ethene
6. (a) 12.5 $dm^3$ (b) 6.75 g
7. (a) 650 $cm^3$ (b) 400 $cm^3$
8. (a) 144 $dm^3$ (b) 10.5 mol
   (c) 1260 $dm^3$
9. (b) (i) 620 $cm^3$ (ii) 330 $cm^3$
   (c) 44, $CO_2$

## Problems involving both Masses of Solids and Volumes of Gases

### Section 4

1. £90 daily
2. (a) 0.30 mol (b) 4.8 $dm^3$
   (c) (i) 1.5 mol (ii) 1.0 mol
   (d) 2.5 g
   (e) 67 g
3. (a) $KO_2$
   (b) $4KO_2 + 2CO_2 \rightarrow 3O_2 + 2K_2CO_3$
   (c) 254 $dm^3$
4. 3.5 g
5. (a) $2H_2O$ on RHS (b) (i) 16.0 g (ii) 33.3 g
6. (a) $2H_2O$ on LHS (b) 1.33 g
7. (a) 4.8 g (b) 1.2 $dm^3$
8. 180 $cm^3$
9. 960 $cm^3$
10. 12.0 $dm^3$

# Chapter 6

## Problems on Concentration

1. (a) 0.0500 mol/$dm^3$
   (b) 1.00 mol/$dm^3$
   (c) 0.250 mol/$dm^3$
   (d) 2.00 mol/$dm^3$
   (e) 0.100 mol/$dm^3$
   (f) 0.125 mol/$dm^3$
   (g) 0.250 mol/$dm^3$
   (h) 0.200 mol/$dm^3$
2. (a) 0.250 mol (b) 0.125 mol
   (c) 0.00500 mol (d) 2.50 mol
   (e) 0.0500 mol (f) 0.0250 mol
   (g) 0.370 mol (h) 1.125 mol

## Problems on Reacting Volumes of Solutions

### Section 1

1. (a) True (b) False
   (c) True (d) False
   (e) True (f) False
2. (a) True (b) False
   (c) True (d) False
   (e) True (f) False
3. (a) False (b) False
   (c) True (d) True
   (e) True (f) False
4. (a) True (b) False
   (c) False (d) False
   (e) True (f) True

### Section 2

1. 1.05 mol/$dm^3$
2. 0.55 mol/$dm^3$
3. 40 $cm^3$
4. (a) $5.0 \times 10^{-3}$ mol (b) $1.0 \times 10^{-2}$ mol
   (c) 0.40 mol/$dm^3$
5. 50 $cm^3$ acid; 0.60 $dm^3$ $CO_2$
6. (a) 2.0 mol/$dm^3$ (b) 400 $cm^3$
7. 33.3 $cm^3$

### Section 3

1. (a) 0.040 g/$dm^3$ (b) $1.9 \times 10^{-4}$ mol/$dm^3$
2. 75%
3. (a) 1.1 mol/$dm^3$ (b) 6.6%
4. 0.145 mol/$dm^3$
5. 84%
6. No, concentration of $K^+ = 5.0 \times 10^{-3}$ mol/$dm^3$
7. Yes, concentration of $Li^+ = 2.0 \times 10^{-3}$ mol/dm
8. (a) 2.4 g (b) 2.4 $dm^3$
9. (b) 3600 $dm^3$
10. (a) $1.5 \times 10^{-6}$ g (b) 1/30
11. (a) 0.050% (b) $2.0 \times 10^{-3}$ mol/$dm^3$
12. 0.109 mol/$dm^3$
13. (a) $1.20 \times 10^{-2}$ mol (b) $1.20 \times 10^{-2}$ mol
    (c) $1.20 \times 10^{-2}$ mol
    (d) 0.276 g (e) 9.2%
14. (a) 50 $cm^3$ (b) 6.6 g
15. 19.2%
16. (a) $2.10 \times 10^{-2}$ mol/$dm^3$
    (b) 1.55 g/$dm^3$
17. (a) $3.00 \times 10^{-3}$ mol/l
    (b) 0.486 g/l

# Chapter 7

## Problems on Heat of Reaction

### Section 1

1. 1380 kJ
2. 12.0 g
3. 85.5 g
4. 45 g
5. 4040 kJ
6. D

### Section 2

1. (a) exothermic
   (b) (i) $55.6 \times 10^3$ kJ (ii) $49.6 \times 10^3$ kJ
2. (a) 12.5 l (b) 230 kJ
   (c) 6.75 g
3. D
4. $CH_4$, 55.6 kJ/g: $C_2H_2$, 50 kJ/g
5. 13.3 kJ
6. $H_2$, 143 kJ/g: $CH_4$, 55.6 kJ/g; C, 32.5 kJ/g
7. (b) (i) −2010 kJ/mol (ii) −3950 kJ/mol
8. (b) 0.1 mol (c) 2 mol/dm$^3$
   (d) 56 kJ (e) 56 kJ
   (f) 112 kJ
9. (b) 30.0 cm$^3$ (c) 2.40 mol/dm$^3$

# Chapter 8

## Problems Reaction Speeds

1. (a) 19 cm$^3$ (b) 30 s
   (c) 44 cm$^3$
2. A
3. C
4. C
5. (b) 0.55 g (c) 10 min
   (d) 2.0 min
6. (a) $2H_2O_2(aq) \rightarrow O_2(g) + 2H_2O(l)$
   (c) 19.7 cm$^3$ (d) 4.3 min
7. (b) (i) 285 s (ii) 65 s
   (c) increase (d) increase
   (e) 20 s

# Chapter 9

## Problems on Solubility

1. 36.0 g/100 g
2. 27 g
3. 4 g of crystals form
4. (a) 50.0 g, 40.0 g, 33.3 g
   (d) 43.0 g/100 cm$^3$ water
5. 62.0 g (19.0 g KCl + 43.0 g $KClO_3$)
6. (a) NaCl (b) 25 g
   (c) 30 g

# Chapter 10

## Problems on Radioactivity

### Section 1

1. (a) 250 c.p.m. (b) 125 c.p.m.
   (c) 62.5 c.p.m.
2. (a) 0.1 g (b) 0.025 g
3. 9 s
4. 48 720 years
5. (a) $^{210}_{83}Bi$ (b) $^{206}_{82}Pb$
   (c) $^{24}_{12}Mg$ (d) $^{14}_{7}X$
   (e) $^{11}_{6}X$ (f) $^{27}_{14}Si + ^{1}_{0}Y$

### Section 2

1. 11 400 years
2. (a) 2000 c.p.m. (b) 250 c.p.m.
3. (b) 120 and 20 c.p.s.
   (c) (i) 33 min (ii) 54 min
   (d) 21 min (e) 5 c.p.s.
   (f) 60 and 10 c.p.s.
   (g) 21 min
4. (a) 250 and 150 c.p.m.
   (b) (i) 115 days (ii) 250 days
   (c) 135 days
5. (b) 1 g

### Section 3

1. (a) 56 years (b) 6.25 g
   (c) 1.56 g (d) 7
2. (a) 11p, 11e, 12n (b) 8p, 8e, 8n
   (c) 25p, 25e, 32n (d) 27p, 27e, 33n
   (e) 82p, 82e, 126n
3. $3.9 \times 10^9$ years
4. A 35 35 115, B 17 17 35, C 47 47 47,
   D 90 142 90, E 27 27 27
5. (a) $^{4}_{2}E$ (b) $^{90}_{37}E$
   (c) $^{146}_{62}E$ (d) $^{234}_{92}E$
6. (i) (a) $^{4}_{2}X$ (b) $^{2}_{1}X$
   (c) $^{0}_{-1}X$ (d) $^{1}_{0}X$
   (ii) (a) $^{4}_{2}He$ (b) $^{2}_{1}H$
   (c) $^{0}_{-1}e$ (d) $^{1}_{0}n$

# List of Approximate Relative Atomic Masses

| Element | Symbol | Atomic number | Relative atomic mass | Element | Symbol | Atomic number | Relative atomic mass |
|---|---|---|---|---|---|---|---|
| Aluminium | Al | 13 | 27 | Lithium | Li | 3 | 7 |
| Antimony | Sb | 51 | 122 | Magnesium | Mg | 12 | 24 |
| Argon | Ar | 18 | 40 | Manganese | Mn | 25 | 55 |
| Arsenic | As | 33 | 75 | Mercury | Hg | 80 | 201 |
| Barium | Ba | 56 | 137 | Neon | Ne | 10 | 20 |
| Beryllium | Be | 4 | 9.0 | Nickel | Ni | 28 | 59 |
| Boron | B | 5 | 11 | Nitrogen | N | 7 | 14 |
| Bromine | Br | 35 | 80 | Oxygen | O | 8 | 16 |
| Cadmium | Cd | 48 | 112 | Phosphorus | P | 15 | 31 |
| Caesium | Cs | 55 | 133 | Platinum | Pt | 78 | 195 |
| Calcium | Ca | 20 | 40 | Potassium | K | 19 | 39 |
| Carbon | C | 6 | 12 | Rubidium | Rb | 37 | 85.5 |
| Cerium | Ce | 58 | 140 | Selenium | Se | 34 | 79 |
| Chlorine | Cl | 17 | 35.5 | Silicon | Si | 14 | 28 |
| Chromium | Cr | 24 | 52 | Silver | Ag | 47 | 108 |
| Cobalt | Co | 27 | 59 | Sodium | Na | 11 | 23 |
| Copper | Cu | 29 | 63.5 | Strontium | Sr | 38 | 87 |
| Fluorine | F | 9 | 19 | Sulphur | S | 16 | 32 |
| Gold | Au | 79 | 197 | Tin | Sn | 50 | 119 |
| Helium | He | 2 | 4 | Titanium | Ti | 22 | 48 |
| Hydrogen | H | 1 | 1 | Tungsten | W | 74 | 184 |
| Iodine | I | 53 | 127 | Uranium | U | 92 | 238 |
| Iron | Fe | 26 | 56 | Vanadium | V | 23 | 51 |
| Krypton | Kr | 36 | 84 | Xenon | Xe | 54 | 131 |
| Lead | Pb | 82 | 207 | Zinc | Zn | 30 | 65 |

# Index